Wireless Networks

Series Editor

Xuemin Sherman Shen, University of Waterloo, Waterloo, ON, Canada

The purpose of Springer's Wireless Networks book series is to establish the state of the art and set the course for future research and development in wireless communication networks. The scope of this series includes not only all aspects of wireless networks (including cellular networks, Wi-Fi, sensor networks, and vehicular networks), but related areas such as cloud computing and big data. The series serves as a central source of references for wireless networks research and development. It aims to publish thorough and cohesive overviews on specific topics in wireless networks, as well as works that are larger in scope than survey articles and that contain more detailed background information. The series also provides coverage of advanced and timely topics worthy of monographs, contributed volumes, textbooks and handbooks.

Minghui Xu • Yifei Zou • Xiuzhen Cheng

Wireless Consensus

Theory and Applications

Springer

Minghui Xu
School of Computer Science and Technology
Shandong University
Qingdao, China

Yifei Zou
School of Computer Science and Technology
Shandong University
Qingdao, China

Xiuzhen Cheng
School of Computer Science and Technology
Shandong University
Qingdao, China

ISSN 2366-1186 ISSN 2366-1445 (electronic)
Wireless Networks
ISBN 978-3-031-70858-9 ISBN 978-3-031-70859-6 (eBook)
https://doi.org/10.1007/978-3-031-70859-6

This Springer imprint is published by the registered company Springer Nature Switzerland AG
The registered company address is: Gewerbestrasse 11, 6330 Cham, Switzerland

Preface

What This Book Is About

In wireless networks, reliable decision-making and coordination are paramount, especially when faced with unreliable communication, harsh environments, and the absence of a central authority. Wireless consensus protocols address these challenges head-on. They enable devices to agree on a common state or course of action, ensuring network coherence and consistency despite these limitations. Furthermore, efficient consensus algorithms are crucial for resource-constrained wireless environments. By minimizing bandwidth and energy consumption, they facilitate scalability and pave the way for widespread adoption. From the Internet of Things (IoT) and mobile computing to distributed robotics, wireless consensus protocols underpin a vast array of critical operations, including coordination and data fusion. With over two decades of active research, wireless consensus has become a cornerstone of modern wireless networks. This book delves deep into this essential topic, providing a comprehensive exploration of various wireless consensus protocols.

The book is divided into five chapters:

Chapter 1: The Power of Agreement: Wireless Consensus Explained. This chapter lays the groundwork by exploring the fundamental concepts of wireless consensus and its historical development. It delves into the key differences between wired and wireless consensus protocols, examining these disparities across the physical layer, network layer, and consensus layer itself. The discussion tackles the unique challenges inherent to wireless consensus, including resource-constrained devices and the impact of unpredictable wireless channels.

Chapter 2: Fault-Tolerant Wireless Consensus. This chapter dives deep into the realm of fault-tolerant wireless consensus. It meticulously details the consequences of unreliable communication and the defining characteristics of fault-tolerant networks. Key concepts like multicast communication, atomic multicast, and the intricacies of designing fault-tolerant consensus mechanisms in wireless environments are introduced. The chapter further explores how failures manifest across different layers and proposes effective methods for handling them, including cross-layer optimization techniques.

Chapter 3: Byzantine Fault-Tolerant Wireless Consensus. This chapter shifts focus to Byzantine fault-tolerant wireless consensus, emphasizing the resilience of consensus protocols against the most severe failures—Byzantine faults. The text delves into various attacks that can cripple such protocols and outlines robust defense strategies at different network layers to counter these attacks.

Chapter 4: Wireless Blockchain. This chapter introduces wireless blockchain as a compelling application of wireless consensus. It explains how blockchain technology seamlessly integrates with wireless networks, unlocking exciting new possibilities. The book provides a foundational overview of blockchain, its specific network requirements, and explores the unique characteristics of wireless blockchain protocols.

Chapter 5: Applications of Wireless Consensus. The concluding chapter explores the diverse applications of wireless consensus. It dissects the specific requirements for both fault-tolerant and Byzantine fault-tolerant wireless consensus, along with the unique considerations for wireless blockchain. To illustrate the practical impact of wireless consensus, the chapter concludes with real-world application scenarios, such as intelligent transportation systems, showcasing the transformative potential of this technology.

This book equips researchers, practitioners, and students with a comprehensive understanding of wireless consensus protocols. It delves into the theoretical foundations while emphasizing practical applications in the context of modern wireless networks.

Qingdao, China

Minghui Xu
Yifei Zou
Xiuzhen Cheng

Acknowledgments

We are deeply grateful to the individuals whose invaluable contributions made this book possible.

Our sincere appreciation goes to Shuo Liu, Li Yang, Guanlin Jing, Xuejian Chi, and Tianyi Sun. Their insightful feedback, dedication, and expertise significantly enriched the content and strengthened the integrity of this publication. Their unwavering commitment and collaborative spirit were instrumental in shaping this book.

Contents

Chapter 1
The Power of Agreement: Wireless Consensus Explained

The importance of wireless consensus in modern wireless communication systems is growing significantly. The proliferation of Internet of Things (IoT) devices and the shift toward edge computing have highlighted the need for effective consensus mechanisms in wireless networks. These networks often operate in decentralized environments where devices must agree on certain decisions without centralized control, making consensus algorithms crucial. The rollout of 5G networks and the ongoing research into 6G have underscored the importance of efficient consensus protocols in wireless communications. These next-generation networks promise higher speeds, lower latency, and increased connectivity, requiring more robust and scalable consensus mechanisms to manage network decisions and resource allocation effectively. With the increasing number of cyber threats, the role of consensus in wireless networks has become more vital in ensuring network security and integrity. Consensus algorithms are integral in maintaining the consistency and reliability of distributed ledger technologies like blockchain, which are increasingly used in wireless network security. The rise of autonomous systems, such as self-driving cars and drones, relies heavily on wireless communication and necessitates reliable consensus mechanisms. These systems require real-time decision-making capabilities, where consensus algorithms play a key role in coordinating actions and information among multiple autonomous units.

This chapter delves into three key aspects of wireless consensus: its evolution, a comparison with wired consensus, and the challenges it faces.

- Section 1.1 explores the history of wireless consensus technologies, tracing their development from early prototypes to their current state-of-the-art implementations.
- Section 1.2 highlights the differences and similarities between wired and wireless consensus mechanisms, examining how the wireless communication medium impacts protocol design and performance.

M. Xu et al., *Wireless Consensus*, Wireless Networks,
https://doi.org/10.1007/978-3-031-70859-6_1

- Section 1.3 discusses the challenges encountered in wireless consensus systems, addressing issues such as limited bandwidth, unreliable communication, and dynamic network topologies.

1.1 Fundamentals of Wireless Consensus

1.1.1 Basic Concepts of Wireless Consensus

Wireless consensus primarily consists of two components: wireless networks and consensus protocols. Wireless networks transmit data through radio waves, offering flexible connectivity and dynamic topology. Consensus protocols ensure that multiple nodes in a distributed system reach an agreement on a specific proposal. Next, we will introduce these two components separately.

1.1.1.1 Wireless Network

Wireless networks can be divided into three parts: wireless communication technology, wireless communication protocols, and wireless network architecture. Each of these components has distinct functions and responsibilities, collectively forming a comprehensive wireless network. Wireless communication technology focuses on the transmission and reception of signals, encompassing various techniques such as modulation, demodulation, and channel coding. Wireless communication protocols define the rules and procedures for communication between devices, ensuring that data can be transmitted efficiently and securely between different devices. Wireless network architecture determines the overall structure and layout of the network, including the distribution of nodes, network topology, and management and allocation of resources. These elements work together to ensure reliability, stability, and efficiency of wireless networks.

As depicted in Fig. 1.1, we summarize three commonly used categories of wireless communication technologies: short-range wireless communication, cellular communication, and low-power wide area network (LPWAN).

Short-Range Wireless Communication Table 1.1 summarizes three short-range wireless communication technologies: Bluetooth, Wi-Fi, and ZigBee. These three communication technologies have different requirements for wireless consensus algorithms due to differences in application scenarios. Bluetooth, because of its short transmission range and fewer connected devices, could be applied in scenarios where wireless consensus and embodied AI are combined. In such scenarios, consensus protocols can enhance cooperation efficiency between embodied AI agents and overall system performance. Wi-Fi, known for its high transmission rate, is suitable for applications with high real-time requirements and low latency, such as wireless consensus among multiple smart devices in a smart home. Zigbee,

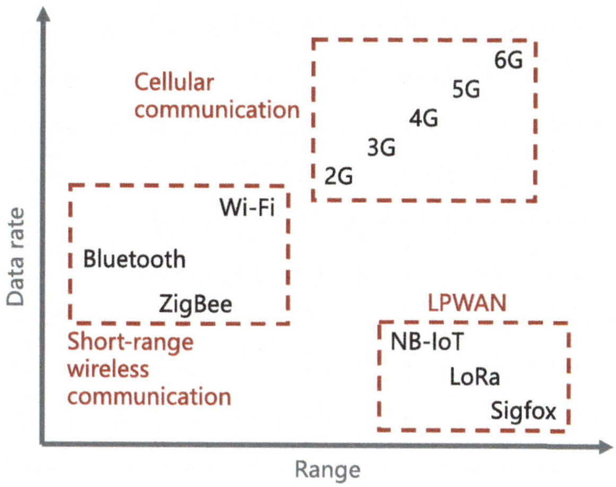

Fig. 1.1 Three wireless communication technologies

Table 1.1 Comparison of Bluetooth, Wi-Fi, and Zigbee [1]

Feature	Bluetooth	Wi-Fi	Zigbee
Standard	IEEE 802.15.1	IEEE 802.11 a/b/g	IEEE 802.15.4
Frequency band	2.4 GHz	2.4/5 GHz	2.4 GHz; 868/915 MHz
Data rate	1 Mbps	54 Mbps	250 Kbps
Range	10 m	100 m	10–100 m
Power consumption	Low	High	Very low
Max number of cell nodes	8	2007	>65,000
Cost	Low	Medium	Low
Applications	Audio, smart devices	Wireless local area network connection	Home automation

with its low power consumption but slower transmission rate, is suitable for scenarios in industries and agriculture where there is a strong demand for low power consumption.

Celluar Communication Cellular communication's biggest advantage compared to the other two communication technologies is its high link reliability due to the need for cell towers to enhance signals. However, cellular communication suffers from lower mobility because of the high and complex deployment cost of these cell towers. While cellular communication can provide a reliable network layer for wireless consensus, it may not fully leverage the advantages of wireless channel broadcasting when treated like a wired connection. Each generation of wireless communication technology, from 2G to 6G, has improved connectivity, speed, and capabilities, enabling a wide range of communication and applications.

Table 1.2 Comparison of Sigfox, LoRa, and NB-IoT [2]

Feature	Sigfox	LoRa	NB-IoT
Standardization	Sigfox company	LoRa-Alliaance	3GPP
Frequency band	Unlicensed ISM bands (868 MHz in Europe, 915 MHz in North America, and 433 MHz in Asia)	Unlicensed ISM bands (868 MHz in Europe, 915 MHz in North America, and 433 MHz in Asia)	Licensed LTE frequency bands
Data rate	100 bps	54 kbps	200 Kbps
Range	10 km (urban), 40 km (rural)	5 km (urban), 20 km (rural)	1 km (urban), 10 km (rural)
Interference immunity	Very high	Very high	Low
Handover	End devices do not join a single base station	End devices do not join a single base station	End devices join a single base station
Applications	Asset tracking	Smart agriculture, smart cities	Healthcare

Table 1.3 Communication primitives

	Base station–based	Base station–less
Routing	OSPF, BGP	AODV, DSDV
Multicasting	PIM, IGMP	ODMRP, MAODV
Broadcasting	802.11 broadcast, LTE broadcast	Flooding, gossiping

Low-Power Wide Area Network Table 1.2 summarizes three common LPWAN technologies: LoRa, narrowband-Internet of Things (NB-IoT), and Sigfox. All three technologies are suitable for long-range, low-power scenarios. Among them, NB-IoT requires cell towers, offers low signal interference, and has relatively higher data transmission rates. Therefore, NB-IoT is better suited for scenarios requiring high link reliability, such as wireless consensus in healthcare. LoRa and Sigfox, both without the need for cell towers and experiencing higher signal interference, differ in terms of data transmission distance and rate. LoRa has a higher data rate than Sigfox, making it more suitable for scenarios with larger data volumes like smart agriculture and smart cities, while Sigfox is better suited for scenarios with smaller data volumes, such as asset tracking.

Routing, multicasting, and broadcasting are essential communication primitives utilized in consensus algorithms. They can be categorized into two types based on the presence or absence of base stations, as shown in Table 1.3.

Routing A routing protocol is a set of rules and procedures used to determine the optimal paths for data transmission across a network, ensuring efficient and reliable delivery of packets from the source to the destination.

Multicasting A multicasting protocol is a communication protocol that enables the transmission of data to multiple specific recipients simultaneously within a network, optimizing resource usage by delivering a single stream of data to multiple locations.

Broadcasting A broadcasting protocol is a communication protocol designed for the distribution of data to all nodes within a network, ensuring that every node receives the same data packet simultaneously, typically used for disseminating information widely and efficiently.

For routing protocols, examples based on base stations include Open Shortest Path First (OSPF) and Border Gateway Protocol (BGP). OSPF is extensively utilized in large enterprise networks and data centers, efficiently routing data through routers to ensure high reliability and optimal performance. BGP plays a crucial role in managing routing between autonomous systems, which is essential for global Internet connectivity. Conversely, examples of base station–less routing protocols such as Ad hoc On-Demand Distance Vector (AODV) and Destination-Sequenced Distance-Vector (DSDV) are designed for mobile ad hoc networks. AODV establishes routes only when needed to adapt to the network's dynamic topology, while DSDV maintains up-to-date route information through periodic updates, making it suitable for small and dynamic networks.

Examples of base station–based multicasting protocols include Protocol Independent Multicast (PIM) and Internet Group Management Protocol (IGMP). PIM is employed in Internet Protocol (IP) networks to efficiently distribute multicast traffic to multiple receivers, supporting applications like video conferencing and streaming services. IGMP manages multicast group memberships within local area networks, facilitating group communication. On the other hand, examples of base station–less multicasting protocols include On-Demand Multicast Routing Protocol (ODMRP) and Multicast Ad Hoc On-Demand Distance Vector (MAODV). ODMRP dynamically establishes and maintains multicast group memberships, often used in military communications and disaster recovery scenarios, while MAODV creates and maintains multicast trees on demand in highly dynamic network environments.

Examples of base station–based broadcasting protocols include 802.11 broadcast and Long-term evolution (LTE) broadcast. The 802.11 protocol supports broadcast communication within local area networks via wireless access points, commonly used in home and enterprise wireless networks. LTE Broadcast delivers content to multiple users simultaneously in mobile networks, ideal for real-time video streaming and broadcast services. Examples of base station–less broadcasting protocols such as flooding and gossiping are employed in networks without fixed infrastructure. Flooding involves each node retransmitting received broadcast messages, making it useful for network discovery and emergency alerts, while gossiping disseminates information by randomly selecting neighboring nodes to propagate messages, making it suitable for large-scale, infrastructure-less networks.

Different network architectures require distinct wireless consensus protocols. For instance, applying single-hop wireless consensus protocols to multi-hop networks can lead to efficiency or security issues. We can describe wireless network architecture from two dimensions: network scale and network dynamics. Based on network scale, wireless networks can be classified into single-hop and multi-hop wireless networks. In single-hop wireless networks, all nodes communicate directly with a central node (e.g., a wireless access point or base station) without the need for intermediate node forwarding. This network structure is simple and suitable for small coverage areas. Typical applications include home wireless networks and public Wi-Fi hotspots. In a home wireless network, there is usually a wireless router to which all devices (such as phones, tablets, and computers) connect directly. In public places like coffee shops and airports, user devices connect directly to the hotspot device providing the wireless service. Multi-hop wireless networks, on the other hand, require data packets to be forwarded through one or more intermediate nodes to reach the destination node. This network structure is more complex and suitable for larger coverage areas. A typical example is a wireless sensor network (WSN), used in fields such as environmental monitoring and agriculture, where sensor nodes transmit data to a sink node via multi-hop communication. Additionally, mobile ad hoc networks (MANETs) are very practical in scenarios like disaster relief and military communications, where mobile devices communicate via multi-hop without the need for fixed infrastructure.

Based on network dynamics, wireless networks can be categorized into static and dynamic wireless networks. In static wireless networks, the positions of the network nodes and the topology remain largely unchanged, resulting in a stable network environment. These networks are commonly found in industrial automation and home automation systems. For example, in a factory's automated control system, the positions of sensors and controllers are fixed, leading to a stable network structure. Similarly, in homes, smart devices like smart bulbs and thermostats communicate through a fixed wireless gateway, which falls under the domain of static wireless networks. Dynamic wireless networks, on the other hand, are highly flexible, with node positions and topologies frequently changing. The network must adapt to these changes to maintain stable communication. A typical example of a dynamic wireless network is a vehicular ad hoc network (VANET), where the movement of vehicles causes constant changes in the network topology, making it suitable for applications in vehicle-to-vehicle communication and intelligent transportation systems. Similarly, drone networks represent another form of dynamic wireless networks, where drones in flight form a network with constantly changing node positions, commonly used in disaster monitoring and logistics delivery.

1.1.1.2 Consensus Protocol

A consensus protocol can be categorized into the following aspects: synchrony abstraction, failure model, and consensus algorithm. Network synchrony describes the temporal characteristics of communication among nodes in the system. The failure model defines the types of faults that nodes in the system might experience. The consensus algorithm is the core mechanism for achieving consistency among nodes in a distributed system.

Synchrony Abstraction The synchrony abstraction is an abstraction of the network layer. This abstraction aims to allow researchers to focus on the design of consensus algorithms without having to overly concern themselves with the complexity and variability of the network layer.

- *Synchronous (Fully Synchronous and Semi-Synchronous):* In synchronous environments, systems operate under a synchronized clock. The delay in message transmission is Δ-bounded, where $\Delta \in [0, 1]$ is a predetermined fixed value. Some systems operate under the strong assumption of *fully synchronous* network channels, which implies no time delay, indicated by $\Delta = 0$. In a *semi-synchronous* model, Δ lies within the interval $(0, 1]$. A synchronous model may be implemented using a universal clock or by leveraging a robust network connection. Moreover, synchronous models facilitate theoretical analysis by providing deterministic time synchronization. However, real-world scenarios are generally asynchronous and at best, partially synchronous.
- *Partially Synchronous:* The partially synchronous model, nestled between the synchronous and asynchronous paradigms, addresses network behaviors with inherent uncertainties [3]. It defines partial synchrony through two scenarios: first, where there exists an undisclosed, definitive upper bound on message transmission delays, known as the a priori bound Δ; second, where this Δ is known but the network's reliability issues allow Δ to be only sporadically effective.
- *Asynchronous:* In the asynchronous model, components of a distributed system operate without the need for a global clock, leading to nonsimultaneous and potentially delayed operations. Under this model, the network assumes that message transmission delays, though finite, are unbounded. The importance of this setting is highlighted by the FLP (named after Fischer, Lynch, and Patterson) impossibility result, which demonstrates that achieving consensus in the event of even a single crash failure is impossible within any bounded period [3]. This result accentuates the inherent challenges of creating reliable distributed systems when faced with unpredictable communication delays and failures.

Failure Model The failure model determines the potential attack methods adversaries may use in the environment where the consensus algorithm is applied. Based on the adversary's capabilities, they can be categorized into six types, with their relationships represented as fail-stop failures $\subset$ crash failures $\subset$ omission failures $\subset$ performance failures$\subset$ authentication-detectable Byzantine failures $\subset$ Byzantine failures.

- **Byzantine Failures.** Nodes exhibit arbitrary and harmful behaviors, such as sending conflicting information or no information at all. For example, a node in a distributed database might falsely report that it has stored data which it actually has not, leading to data loss.
- **Authentication-Detectable Byzantine Failures.** Nodes act arbitrarily but cannot alter verifiable information due to enforced authentication. For instance, a node might send a false update request but is caught because it cannot forge the required digital signature.
- **Performance Failures.** Messages are sent and received correctly but at inappropriate times, breaching the protocol's timing requirements. An example of this is a node sending a time-critical vote for consensus after the deadline has passed, which causes a delay in the decision-making process.

- **Omission Failures.** Nodes fail to respond or act, missing the necessary communication windows. An example would be a node failing to acknowledge a coordination message, leading to a timeout in the consensus process.
- **Crash Failures.** A node stops operating and becomes unresponsive due to internal errors or resource exhaustion. For example, a server hosting a node might overheat and shut down unexpectedly, ceasing all communications.
- **Fail-Stop Failures.** A node crashes and its failure is immediately known to other nodes through predefined detection mechanisms. For example, a node might lose power and send a last-gasp message alerting others to its imminent shutdown.

Consensus Algorithm Consensus algorithms are the core of consensus protocols, determining how to achieve consensus. These algorithms define the processes and rules that ensure all participating nodes agree on a single data value or state, despite the presence of faults or inconsistencies in the network. Consensus algorithms can be broadly categorized into fault-tolerant consensus, Byzantine fault-tolerant consensus, and Proof-of-X consensus. Fault-tolerant consensus algorithms handle non-malicious failures, ensuring agreement. Byzantine fault-tolerant consensus algorithms address malicious or arbitrary faults, providing robustness against adversarial attacks. Proof-of-X consensus algorithms, such as Proof-of-Work, Proof-of-Stake, and Proof-of-Authority, rely on various proof mechanisms to achieve consensus, commonly used in blockchain and cryptocurrency applications to maintain a decentralized ledger.

Definition 1.1 (Consensus) There are n nodes, of which at most f might crash, i.e., at least $n - f$ nodes are correct. Node i starts with an input value v_i. The nodes must decide for one of those values, satisfying the following properties:

- **Agreement** *All correct nodes decide for the same value.*
- **Validity** *The decision value must be the input value of a node.*
- **Termination** *All correct nodes terminate in finite time.*

Paxos is a renowned consensus algorithm developed by Leslie Lamport in the 1990s, designed to achieve consistency among multiple replicas in a network, especially within fault-prone distributed systems. The Paxos algorithm ensures that multiple replicas within a network can reach an agreement on certain values (such as the system's next state), even in the face of challenges such as message loss and replica failures. The basic process of the Paxos (as shown in Fig. 1.2) algorithm is divided into three main stages:

- ***Prepare Phase:*** In the prepare phase, a replica, acting as the proposer, attempts to become the leader by sending a "prepare request" with a unique, highest proposal number to other replicas called replicas. If the proposal number is higher than any number the replicas have previously committed to, they promise not to accept any new proposals with lower numbers and return information about the highest-numbered proposal they have previously accepted (if any) to the proposer.
- ***Propose Phase:*** During the propose phase, the proposer decides on the value of the proposal based on feedback from the replicas, prioritizing the value of

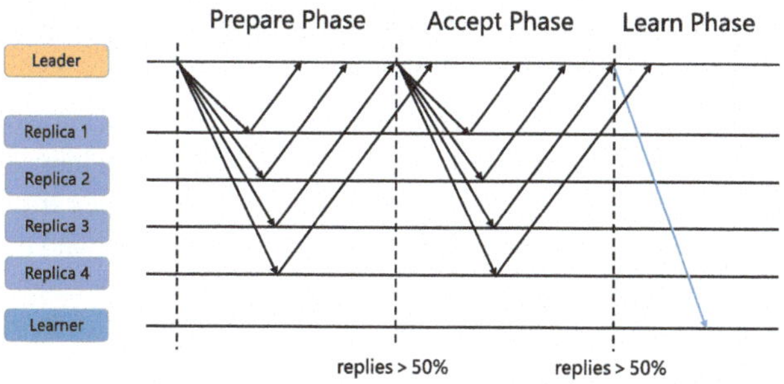

Fig. 1.2 Paxos

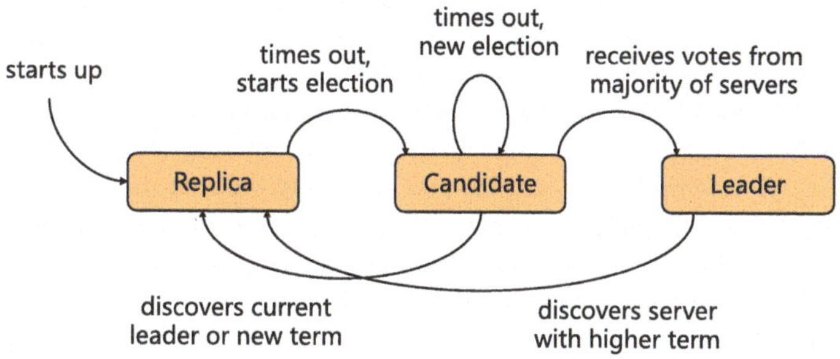

Fig. 1.3 RAFT

the highest-numbered proposal received. If no accepted proposal values are available, the proposer may choose a new value. The proposer then sends a "proposal" containing this value to all replicas. Upon receiving the proposal, if its number is not less than the lowest number they have promised, the replicas accept this proposal and send an acceptance notification back to the proposer.

- ***Learn Phase:*** In the learn phase, once the proposer receives acceptance notifications from the majority of replicas for the same proposal, the proposed value is confirmed as selected. The proposer then broadcasts this result to all replicas (including other proposers and learners) so that all participants can update their status and record this consensus result.

Through its rigorous coordination of proposers, replicas, and learners, Paxos exemplifies a pivotal methodology in consensus algorithms, proving essential for the integrity and reliability of modern distributed architectures. Its pivotal role in shaping the landscape of distributed computing illustrates not only its practical utility but also its theoretical importance, making it a cornerstone in the field of distributed systems.

RAFT (as shown in Fig. 1.3) is designed to manage log consistency in distributed systems, proposed by Diego Ongaro and John Ousterhout in 2014. The goal of RAFT is to provide the same functionalities as algorithms like Paxos, but with a design that is easier to understand. The RAFT algorithm breaks down the consensus problem into the following three key subproblems:

- ***Leader Election:*** RAFT requires a leader to manage the replication of log entries. Therefore, the primary task is to elect a leader. This is achieved through an election process where a replica, upon not receiving messages from the current leader for a sufficiently long period, increases its term number and transitions to a candidate state, then requests votes from other replicas. If the candidate receives votes from at least $\frac{N}{2} + 1$ replicas in the cluster, it is elected as the new leader.
- ***Log Replication:*** Once a leader is elected, the leader replica is responsible for handling client requests, each including a new operation. The leader first appends the operation as a new log entry to its log and then replicates this log entry to other replicas. The operation is only committed and applied to the state machine of the cluster once the log entry has been replicated to the majority of replicas and confirmed by the leader.
- ***Safety:*** To ensure the correctness and consistency of the consensus, RAFT introduces several safety measures. These include ensuring that the elected leader possesses all committed log entries and enforcing the rule that if two logs have the same entry at a particular index, then all preceding entries in those logs are also identical. These rules help the system maintain consistency and prevent data loss in the face of network partitions and replica failures.

The solutions to these three subproblems constitute the core mechanisms of RAFT, making it an easy-to-understand and effectively implemented consensus algorithm for distributed systems. Additionally, the RAFT algorithm includes mechanisms for Leader Change. This process may be initiated either actively or passively: leaders can proactively transfer their roles due to planned maintenance or upgrades, or a new election can be triggered automatically if the current leader becomes unreachable due to failures or network issues. This flexibility ensures that the cluster can maintain its operational continuity without a leader for only a minimal amount of time.

Sync HotStuff In 2020, Abraham et al. [4] introduced Sync HotStuff, an algorithm designed to enhance Byzantine fault tolerance in synchronous distributed systems. Sync HotStuff addresses the limitations of earlier synchronous protocols by simplifying consensus achievement and ensuring robust fault tolerance. It consists of a two-phase, leader-based approach that tolerates up to half of the participants being Byzantine faults. The primary actors in the algorithm are the leader, who proposes blocks, and the replicas, who vote on these proposals, ensuring that the system operates effectively even under weaker, more realistic network synchrony conditions. This methodological refinement significantly improves both the usability and reliability of systems in environments such as financial services and blockchain technology.

Sync HotStuff (as shown in Fig. 1.4) is structured around several essential processes:

- ***Propose:*** In the standard synchrony model, the leader broadcasts a proposal which the replicas then echo back. If the leader fails to propose, the replicas

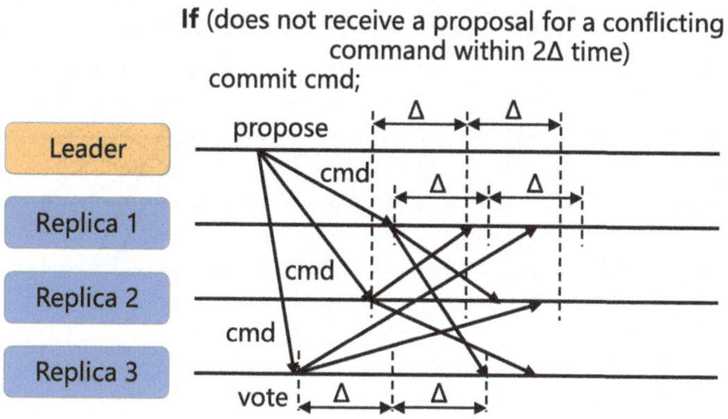

Fig. 1.4 Sync HotStuff

will timeout and initiate a leader change. This step ensures that all replicas have received the proposal and are ready to proceed to the voting stage.

- ***Vote:*** Upon receiving a proposal, replicas will vote for it unless they receive a conflicting proposal signed by the leader during the maximum round-trip delay. This voting step ensures that there is consensus among the replicas about the proposed block before moving to the pre-commit phase. If a replica detects equivocation from the leader (i.e., conflicting proposals), it will not proceed with voting and instead initiate a leader change.
- ***Pre-Commit:*** After receiving sufficient messages in the form of $\langle \text{propose}, Bk + 1, v, Cv(Bk)\rangle_L$ from other replicas, indicating that the next block in sequence has been proposed and that the current block Bk is unchallenged, a replica will pre-commit the block Bk and broadcast a commit message $\langle \text{commit}, Bk, v\rangle_r$. This step prepares the network for the final commitment of the block, assuming no conflicts are detected.
- ***Commit:*** Upon receiving sufficient commit messages in the form of $\langle \text{commit}, Bk, v\rangle$ from other replicas, the replica commits block Bk and all its ancestor blocks. This final commitment ensures that the block and its predecessors are permanently added to the blockchain, confirming the transactions contained within.

Sync HotStuff introduces practical improvements over previous Byzantine fault-tolerant (BFT) mechanisms by simplifying the complexity associated with synchronous protocols and leveraging a realistic approach to network delays and faults. Its deployment demonstrates a robust and scalable solution suitable for high-stakes distributed systems, effectively bridging theoretical resilience with practical deployment needs.

PBFT In 1999, Miguel Castro and Barbara Liskov [5] introduced a novel algorithm for addressing Byzantine faults in asynchronous systems, termed practical Byzantine fault tolerance (PBFT). Developed to counteract the arbitrary and potentially

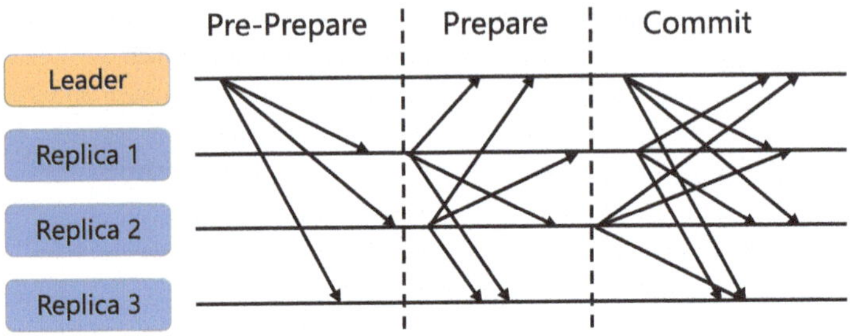

Fig. 1.5 PBFT

malicious behavior of system components, this algorithm enhances the reliability and integrity of distributed computing environments. It is structured around a cohesive framework involving multiple participants, including clients, a primary replica, and several backup replicas. The algorithm is divided into key components that manage request processing, consensus on system states, and execution of operations in a fault-tolerant manner.

The PBFT algorithm (as shown in Fig. 1.5) is designed to operate within an asynchronous network environment, recognizing the inherent unreliability of such systems where messages may be delayed, lost, duplicated, or delivered out of order. To accommodate the potential for arbitrary failures, the algorithm adopts a Byzantine failure model, which allows for the assumption that any system component might fail and behave in unpredictable ways due to malicious attacks or software errors. This setup prepares the system to robustly handle the unpredictable nature of distributed network communications and replica behavior.

The process of the PBFT algorithm is as follows:

- ***Pre-Prepare Phase:*** The primary replica receives a request from a client and assigns it a sequence number. It then multicasts a *pre-prepare* message to all backup replicas, which includes the request's digest but not the entire request, to keep the message size small. This initiates the consensus process.
- ***Prepare Phase:*** Upon receiving the *pre-prepare* message, each backup replica validates it and multicasts a *prepare* message to all other replicas. A replica transitions to the *prepared* state for a request if it receives *prepare* messages from at least $\lceil \frac{2N}{3} \rceil$ (to ensure the resilience of the system against up to $\frac{N-1}{3}$ Byzantine replicas) distinct replicas, where N is the total number of replicas. This phase ensures agreement on the request's sequence number and content.
- ***Commit Phase:*** Once a replica has received *prepare* messages from the majority as specified above, confirming that a quorum of replicas agrees with the request order and content, it multicasts a *commit* message. The replica commits the request locally once it receives *commit* messages from at least $\lceil \frac{2N}{3} \rceil$ distinct replicas. This ensures that all non-faulty replicas execute requests in the same sequence.

- ***Reply Phase:*** After the commit, each replica executes the request and sends a reply to the client. This phase guarantees that all replicas process requests consistently and in the same order.
- ***Checkpoint and Log Truncation:*** Periodically, the replicas engage in a checkpointing process where they agree on a stable state of the system. This process involves receiving confirmation from at least $\lceil \frac{2N}{3} \rceil$ replicas. It allows for the truncation of the log, which reduces the amount of stored data and aids in system recovery from crashes.
- ***View Change Protocol:*** This mechanism addresses situations where the primary may be faulty. It involves a new election for the primary through a consensus process similar to the initial phases, requiring confirmation from at least $\lceil \frac{2N}{3} \rceil$ replicas to establish a new primary and ensure the system's continuous operation.

The introduction of the PBFT algorithm by Castro and Liskov marked a significant advancement in the field of distributed systems, offering a robust solution for maintaining system continuity despite Byzantine failures. Its implementation within an NFS system demonstrated minimal impact on performance, affirming its practicality for real-world applications. This algorithm not only fortifies the resilience of distributed architectures but also serves as a foundational model for subsequent research in fault tolerance.

HoneyBadgerBFT In 2016, Andrew Miller et al. [6] proposed the HoneyBadgerBFT algorithm, an innovative Byzantine fault-tolerant (BFT) consensus protocol designed for asynchronous environments typical in decentralized systems like cryptocurrencies. Unlike traditional BFT protocols that rely on timing assumptions for network synchronization, HoneyBadgerBFT (as shown in Fig. 1.6) operates independently of any timing conditions, ensuring robustness and liveness even in the most adverse network conditions. The key innovation of HoneyBadgerBFT lies in its ability to achieve consensus without the need for network timing assumptions, making it highly effective and secure in unpredictable network environments.

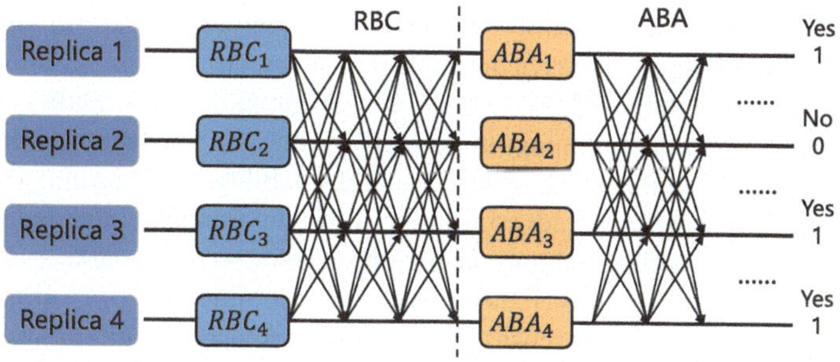

Fig. 1.6 HoneyBadger BFT

The process of the HoneyBadgerBFT algorithm is as follows:

- ***Initialization and Setup:*** Each replica is initialized with a public key for Threshold Public-Key Encryption (TPKE), which is vital for the secure encryption and decryption of transactions. This setup phase is pivotal in ensuring the confidentiality of transactions prior to reaching a consensus.
- ***Transaction Input and Encryption:*** Replicas receive transactions as inputs. In preparation for consensus, each replica selects a subset of transactions from its buffer and encrypts them using the TPKE scheme. This step safeguards the confidentiality of the proposed transactions until a consensus is achieved.
- ***Execution of the Asynchronous Common Subset (ACS) Protocol:*** Encrypted transactions are input into the ACS protocol, which orchestrates coordination between Reliable Broadcast and Byzantine Agreement protocols. This ensures that all honest replicas concur on a common set of encrypted transactions. The Reliable Broadcast mandates that each message be acknowledged by at least $N - f$ replicas (where N is the total number of replicas and f is the number of permissible faulty replicas), guaranteeing that a quorum of replicas receives each transaction.
- ***Decryption and Agreement:*** Upon reaching a consensus on a common subset of transactions, replicas collaborate using their decryption shares to decrypt the transactions. This phase requires at least $f + 1$ decryption shares to decrypt a transaction successfully, thereby ensuring the integrity and security of the decryption process.
- ***Transaction Ordering and Finalization:*** Following decryption, transactions are ordered and finalized according to the sequence established during the ACS protocol. This ensures a consistent and tamper-proof ledger, adhering to the agreed-upon order of transactions.
- ***Committing Transactions to the Blockchain:*** Finally, the ordered transactions are committed to the blockchain, thereby updating the system's state based on the validated and agreed-upon transactions. This final step signifies the completion of the consensus process and the integration of new transactions into the blockchain.

HoneyBadgerBFT significantly enhances the resilience and throughput of blockchain technologies in asynchronous networks. Its design eliminates the dependency on network delays, which are a common vulnerability in other BFT protocols, thereby aligning it perfectly with the needs of decentralized applications that require high security and availability. The protocol's ability to operate efficiently in an adversarial network environment and handle high transaction volumes without compromising on speed or security marks a substantial advancement in the field of distributed consensus algorithms.

The Proof-of-X (PoX) consensus encompasses protocols in the blockchain for selecting a leader replica to propose a block. Nodes use resources, either physical–like mining hardware or RAM–or virtual–such as stakes or reputation, to qualify as leaders. Figure 1.7 illustrates the two PoX categories. Nodes compete by demonstrating resource usage, gaining the chance to propose blocks.

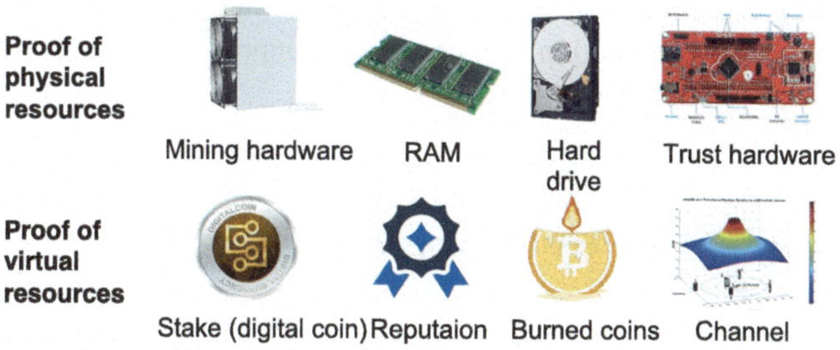

Fig. 1.7 Proof-of-X consensus protocols

In PoW-based consensus algorithms, miners compete with each other to add new blocks to the blockchain using their computational power. A miner with more computational power has a higher probability of proposing the next block. Bitcoin is the first global decentralized transaction ledger that employs a PoW consensus algorithm based on the hashcash with double-iterated SHA-256. Ethereum makes use of a PoW algorithm called Ethash, which is rooted in the Dagger-Hashimoto algorithm and designed to be memory-hard. This means that finding a nonce requires a lot of memory and bandwidth, making it difficult for parallel computing. To address the scalability issues in Bitcoin, Bitcoin-NG separates Bitcoin's PoW-based consensus process into two subprocesses: leader election and transaction serialization, with the former still utilizing a PoW-based consensus algorithm to agree on a key block, whose proposer is responsible for the next epoch.

Proof-of-Work (PoW) involves miners using computational power to compete for block addition. The more computational power, the higher the chance of becoming the next proposer. Bitcoin employs a PoW algorithm based on hashcash with double-iterated SHA-256, while Ethereum uses Ethash, a memory-hard algorithm derived from Dagger-Hashimoto. Bitcoin-NG addresses scalability by splitting the PoW process into leader election and transaction serialization. Other physical resource–based proofs include Proof-of-Space (PoSpace) [7], Proof-of-Elapsed-Time (PoET) [8], and Proof-of-Retrievability [9], with applications in decentralized storage networks and cloud servers. PoSpace is common in networks like Filecoin, Storj, and Sia. PoET, developed by Intel using SGX, involves a random wait time enforced by trusted hardware. Permacoin [10] combines storage and computational power for consensus.

Due to PoW's high energy consumption, Proof-of-Stake (PoS) emerged as an efficient alternative, where block creators are chosen based on their stakes in the network. Variants like Delegated Proof-of-Stake (DPoS) [11], Proof-of-Authority (PoA) [12], and others including Proof-of-Activity [13], Proof-of-Reputation (PoR) [14], Proof-of-Burn (PoB) [15], and Proof-of-Channel (PoC) [16–18] further diversify the proof concepts. DPoS optimizes validation efficiency by electing delegates responsible for transaction validation and network maintenance. PoA, designed for permissioned blockchains, permits only approved validators to propose blocks, while PoR enhances security by using reputation scores. PoB involves coin destruction to earn the right to propose blocks, used by Slimcoin [19]. PoC incorporates channel competition into consensus, improving efficiency in wireless networks under adversarial conditions.

1.1.2 History of Wireless Consensus

The evolution of wireless consensus algorithms follows a pattern seen in the development of many other algorithms. Initially, these algorithms were predominantly confined to theoretical explorations, serving as conceptual frameworks without immediate practical applications. However, as wireless communication technologies advanced at a rapid pace, the importance of wireless consensus mechanisms became increasingly apparent. Researchers recognized the potential impact of these protocols on real-world wireless networks and shifted their focus toward practical implementations. They began deploying wireless consensus protocols in actual network environments, rigorously testing their performance and identifying areas for improvement. This practical approach aimed to optimize the efficiency of wireless consensus in real-world systems, ensuring that the protocols could meet the demands of modern wireless communication infrastructures.

1.1.2.1 Theoretical Level

The theoretical exploration of consensus algorithms underwent two distinct phases. Initially, researchers aimed to develop wireless consensus theories applicable to all types of wireless networks, encapsulating them through an abstract MAC layer. Subsequently, it became evident that different wireless networks exhibited distinct characteristics, leading to the emergence of various branches focused on specific types of wireless networks at the theoretical level.

I. Consensus Algorithms from Abstract MAC Layer

The study of wireless networks encompasses a diverse array of communication considerations, such as slot-based operations, collision handling, and signal-to-noise ratio considerations. These complexities introduce significant variability in the assumptions researchers must make, complicating the development of universal algorithms and the comparison of different network behaviors. The abstract MAC layer [20] provides a uniform way to describe local broadcast communication, which simplifies the complexity of dealing with specific channel behaviors. The abstract MAC layer (as shown in Fig. 1.8) is a theoretical construct that simplifies the design of algorithms by providing reliable local broadcast without detailing the underlying channel model. It offers timing guarantees based on abstract delay functions. These functions are influenced by the network's level of contention. These functions enable researchers to analyze algorithm performance, which is independent of the specific behaviors of the network. The abstract MAC layer

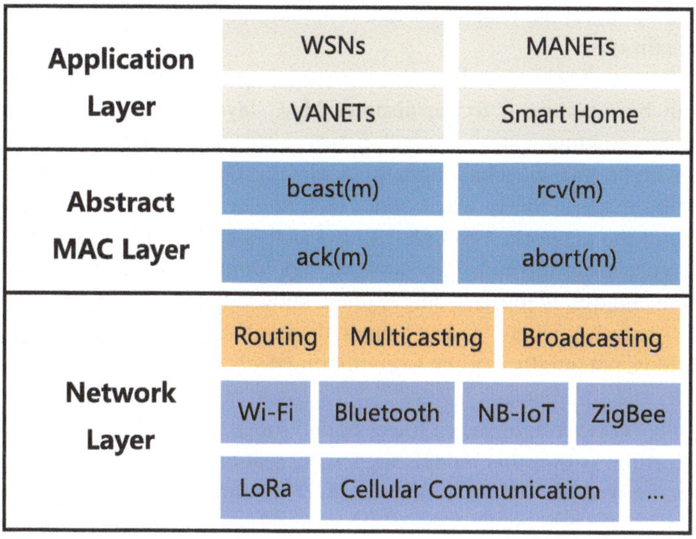

Fig. 1.8 Abstract MAC layer

provides a powerful abstraction for studying wireless networks, enabling the development of robust algorithms.

Based on the abstract MAC layer, researchers have conducted extensive theoretical studies on wireless consensus. These studies can be broadly categorized into several types: one group of research focuses on the upper and lower bounds of wireless consensus algorithms, attempting to define the performance limits under the best and worst conditions. Another group examines the impact of network scale and whether node IDs are known on wireless consensus, as these factors play a crucial role in the design and efficiency of the algorithms. As wireless technology continues to advance, the focus of research has gradually shifted from theoretical exploration to practical application. Some researchers are now investigating how to deploy the abstract MAC layer in real-world scenarios to achieve efficient communication primitives.

Upper and Lower Bounds The upper and lower bounds of wireless consensus algorithms are critical metrics for evaluating their performance and efficiency. The upper bound typically refers to the best performance that the algorithm can achieve under ideal conditions, such as maximum data transmission speed, node synchronization time, and energy consumption. These upper bounds provide a benchmark, indicating the potential of the wireless consensus algorithm under optimal conditions. Conversely, the lower bound refers to the minimum performance level that the algorithm can maintain under the worst-case scenarios. This includes the lowest possible transmission speed, synchronization accuracy, and energy efficiency in the presence of severe network delays, maximum channel interference, or the highest frequency of node failures. Understanding these lower bounds is crucial as they reveal the robustness and reliability of the algorithm under the most adverse conditions.

Newport et al. [21] expand on the classical impossibility of consensus with a single fault by adapting it to the abstract MAC layer model, which reflects the asynchronous communication characteristics inherent in wireless networks. This adaptation highlights the challenges of achieving deterministic consensus under such conditions, particularly when nodes lack unique identifiers or comprehensive knowledge of the network. Through detailed analysis, the authors demonstrate that achieving consensus in multi-hop network topologies is unfeasible unless nodes possess both unique identifiers and explicit knowledge of the network size. This is due to scenarios where nodes cannot distinguish between different network configurations, potentially leading to conflicting decisions that breach the consensus agreement.

The study further establishes a lower bound for the time complexity of any consensus algorithm within this model, suggesting that effectively reaching consensus requires at least $\Omega(D \cdot F_{ack})$ time, where D denotes the network diameter and F_{ack} represents the maximum delay in message acknowledgment. This lower bound encapsulates the minimum time required for information to traverse the network, accounting for the unpredictable behavior typical of the wireless medium and the mechanisms of the abstract MAC layer.

Moreover, Newport et al. [21] present two algorithms that not only meet but precisely match these lower bounds, thereby demonstrating their optimality. The first, designed for single-hop networks, achieves consensus in an optimal time of $O(F_{ack})$, aligning with the upper bound dictated by message acknowledgment delays. The second algorithm, wPAXOS, is optimized for multihop networks and attains consensus in $O(D \cdot F_{ack})$ time, efficiently leveraging the network structure and acknowledgment timings to expedite the consensus process. This approach ensures adherence to the established upper time complexity bound, illustrating a seamless integration of theoretical limits and practical algorithmic solutions in wireless network consensus.

Network Size In practical applications, precisely determining the exact number of participants and the overall scale of the network is often infeasible. This challenge arises from the dynamic and distributed nature of wireless networks, where the number of nodes can change at any time, and the joining or leaving of nodes is unpredictable. Additionally, the coverage area of the wireless network and the geographical locations of participants may also vary, further complicating real-time monitoring of the network scale.

Addressing these real-world constraints, Newport et al. [22] have built upon and expanded their prior research [21], introducing the first known fault-tolerant consensus algorithms suitable for the abstract MAC layer model. They introduced two innovative randomized algorithms that address the fault-tolerant consensus challenges within this model.

The first algorithm known as *Counter Racing* eliminates the need for preexisting knowledge about network size or the identities of its participants. It impressively ensures that consensus is achieved with high probability within a polynomial timeframe, even in the face of numerous faults. This robust approach is particularly notable for its ability to maintain network coherence and reliability under a broad spectrum of conditions. The second algorithm, *Randomized Delays*, opts for a trade-off by slightly relaxing certain consensus properties to enhance the termination state. This strategic compromise allows the majority of nodes to converge on a unified value more swiftly. The termination time for this algorithm is significantly faster than that of the Counter Racing algorithm–it is a linear factor of the latter's termination period but achieves this without significant delays. They verify the performance and correctness of the proposed algorithms through theoretical analysis rather than experimental simulation.

Implementing Abstract MAC Layer Deploying an abstract MAC layer in practical applications can significantly simplify the design of higher-layer wireless consensus algorithms. The abstract MAC layer provides standardized communication interfaces and basic communication primitives, allowing developers to focus on high-level logic and optimization without dealing with complex lower-layer details. An efficient abstract MAC layer can provide reliable communication services, reducing the impact of network fluctuations and interference and thereby enhancing the overall performance of wireless consensus algorithms. In summary, the effective implementation of the abstract MAC layer significantly improves the development efficiency and system performance of wireless consensus algorithms, driving advancements in wireless communication technology.

Yu et al. [23] address the problem of the exact implementation of the abstract MAC layer (absMAC) in the physical SINR model. They propose a distributed algorithm to implement the absMAC layer, aiming to minimize the bounds of the abstract delay functions within a specific communication model that defines concrete channel behaviors. The algorithm utilizes physical carrier sensing, a common feature in wireless devices, to achieve an efficient and exact implementation.

Specifically, they present an algorithm that exactly implements the absMAC layer, providing asymptotically optimal bounds for both the acknowledgment and progress functions defined in the absMAC layer. This algorithm leads to the development of faster solutions for high-level problems in the SINR model, as demonstrated by the algorithms for consensus, multi-message broadcast, and single-message broadcast. The implementation algorithm is based on an optimal solution to the general local broadcast (GLB) problem, which, for the first time, considers the number of distinct messages. The algorithm improves delay bounds and requires less global information compared to previous work. For the first time, the study incorporates the number of distinct messages as a parameter in the definition of the local broadcast problem, allowing the algorithm to handle a wider range of communication scenarios.

They verified the algorithm's correctness through theoretical analyses and constructive proofs. They conducted simulations for the absMAC layer implementation, comparing its performance with existing algorithms. They evaluated the impact of network density and the parameter ϵ on the algorithm's performance. For the

GLB algorithm, they evaluated the impact of network size n, the number of distinct messages k, and the model parameter ϵ on the algorithm's performance.

II. Specific Wireless Network

Subsequently, researchers began investigating consensus protocols tailored to the specific types of wireless networks. This phase marked a shift in consensus algorithm research toward practical systems. The primary focus during this period was on four types of wireless network characteristics: collision detectors, multipath fading, probabilistic broadcast, and ad hoc networks.

Collision Detectors Collision detectors are fundamental elements in wireless ad hoc networks, designed to mitigate erratic message losses due to collisions, electromagnetic interference, and other communication anomalies. These detectors constantly monitor the broadcasting medium to detect when messages are lost due to collisions and issue notifications, thus preserving the integrity of data transmission. However, they do not provide specifics regarding the number of messages lost or the identities of their senders.

In wireless networks, it is not necessarily guaranteed that a node transmitting a message can detect collisions, setting these networks apart from others, such as Ethernet networks. To more effectively categorize and utilize collision detectors, researchers classify these detectors by "completeness"–the ability to detect actual collisions–and "accuracy," which ensures that only actual collisions are reported, avoiding false positives. Completeness indicates how effectively a collision detector identifies collision occurrences, while accuracy focuses on the detector's ability to avoid false alarms, thus reducing unnecessary complexity.

Chockler et al. [24] systematically propose a set of consensus algorithms for single-hop environments in wireless ad hoc networks, particularly where nodes are prone to crashes, with each node within the broadcasting range of others. They address the core challenge of unpredictable message loss due to collisions, electromagnetic interference, and other issues. They introduced a major breakthrough with a new classification scheme for collision detectors based on "completeness" and "accuracy." Their classification shows how to solve the consensus problem with support from various types of collision detectors and provides the corresponding lower bounds for complexity. Their approach not only offers theoretical novelty but also guides the design of consensus algorithms, enhancing fault tolerance and adaptability through dynamic adjustments based on network conditions.

To verify the effectiveness of the algorithms they proposed, they use simulation techniques. Using simulations with a realistic 802.11 MAC layer implementation and a detailed radio propagation model, they confirm the performance of the algorithms under various node densities and MAC layer parameters. They demonstrate the algorithms' performance in single-hop networks and use the single-hop consensus algorithm to solve the multi-hop network consensus problem, showcasing the algorithms' robustness in challenging and noisy environments. Furthermore, they display experimental results that show how the algorithms perform under different node densities and changing network conditions. These experiments confirm the theoretical properties of their proposed algorithms and demonstrate their potential for application in real network scenarios, especially regarding stability and efficiency when network density and communication interference conditions vary.

Multipath Fading Multipath fading is a phenomenon in wireless communications where signals transmitted from a source are reflected by various objects, such as buildings, trees, and the ground. These reflections create multiple signal paths that travel to the receiver. Consequently, the receiver captures multiple delayed versions of the same signal, arriving at different times and angles. These varying signal paths lead to both constructive and destructive interference, causing fluctuations in the amplitude, phase, and path length of the received signal. Such fluctuations can significantly degrade the communication system's performance by inducing rapid changes in signal strength and potentially causing temporary information loss. The extent of multipath fading is influenced by the environment's structure, the signal's frequency, and the mobility of the transmitting and receiving devices. Techniques like diversity reception and channel equalization are commonly used to mitigate the negative effects of multipath fading and to ensure reliable communication across wireless channels.

Scutari et al. [25] address the challenges of achieving distributed consensus in wireless sensor networks impacted by multipath fading. They note that conventional consensus algorithms typically overlook the complex dynamics introduced by propagation delays and multipath effects, which can significantly affect the network's ability to reach a common decision. They have developed a consensus algorithm that adaptively adjusts to these conditions, ensuring that distributed nodes can reliably converge to a globally optimal decision statistic despite the uncertainties introduced by multipath fading.

The innovative essence of their research resides in establishing the necessary and sufficient conditions for the proposed algorithm to converge exponentially to the optimal decision statistic across all scenarios of bounded delay. By modeling each network link as a multipath, frequency-selective channel, they have rendered their model exceptionally realistic and directly applicable to real-world wireless communications. They demonstrate that the convergence rate of the algorithm is influenced by channel parameters and propagation delays, offering essential insights for the design and optimization of sensor networks functioning in environments with multipath interference.

To validate their theoretical advancements, they undertook extensive numerical simulations. These simulations showcased the algorithm's robustness across diverse network scenarios, including those with severe propagation delays and asymmetric channel conditions. The findings reveal that the algorithm can secure globally optimal estimates without directly estimating channel coefficients. This methodology streamlines the implementation in practical sensor networks and underscores the algorithm's capacity to adapt to various challenging communication environments.

Probabilistic Broadcast Probabilistic broadcasting is a technique used in wireless networks to improve the efficiency of achieving consensus among nodes. In this method, nodes wake up randomly and broadcast their current states to neighboring nodes. Those within a specific reception radius receive this broadcast and update their states based on the received information. This approach contrasts with traditional consensus algorithms, which typically require establishing a reliable bi-directional communication link—a process that can be inefficient and susceptible to collisions in wireless environments. Probabilistic broadcasting utilizes the natural broadcasting capabilities of wireless media, allowing nodes to asynchronously update their states based on the likelihood of receiving a broadcast, influenced by factors such as distance and signal strength.

Aysal et al. [26] examine the impact of probabilistic broadcasting on the speed and efficiency of consensus achievement in wireless networks. This discussion extends a class of non-sum-preserving algorithms that employ unidirectional broadcasting to accommodate a time-varying connection model. The principal innovation is the demonstration that these algorithms almost surely converge to a consensus, which represents the true average of the initial node values. They provide a straightforward bound on the convergence time to validate this approach, revealing that for specific network configurations and fading models, the rate of convergence can be remarkably fast. The effects of various fading models on the performance of the consensus protocol are analyzed. For instance, a lognormal shadowing model is considered to understand how signal-to-noise ratios at different nodes influence reception probabilities and, consequently, overall convergence.

Furthermore, a novel matrix update methodology is introduced to describe the state changes within the network. This involves the calculation of expectations for various matrix functions that describe the network's state transitions. These mathematical formulations assist in bounding the convergence time using the largest eigenvalues of specific averaging matrices derived from the network's connectivity and node update probabilities.

In terms of experiments, they provide simulations for networks of moderate size to compare the efficiency of probabilistic broadcasting with that of traditional gossip algorithms. These simulations show that probabilistic broadcasting can achieve faster consensus, especially in networks of moderate size under specific conditions governing the broadcast probability.

Ad Hoc Network Multi-hop ad hoc networks are decentralized and flexible wireless networks that operate without relying on any preestablished infrastructure. Within these networks, each node functions as both a host and a router, forwarding packets to facilitate communication across multiple wireless hops. This architecture enables nodes outside each other's direct communication range to connect via intermediate nodes. These networks are inherently self-configuring and self-healing; they automatically establish and maintain network connectivity as nodes dynamically move or adjust their positions. This makes them especially valuable in situations where setting up fixed infrastructure is impractical or prohibitively expensive, such as in disaster recovery operations, military battlefield communications, and spontaneous or temporary events.

In their research, Borran et al. [27] explore how consensus algorithms, especially the Paxos/LastVoting algorithm, can be adapted for multi-hop ad hoc networks. They pinpoint the limitations of traditional consensus algorithms, which are mainly developed for wired networks, and find these algorithms underperform in wireless ad hoc networks due to their inability to effectively handle link failures. To address this issue, they refine the Paxos/LastVoting algorithm with a robust communication layer that is specifically designed to cope with the unpredictable dynamics of wireless ad hoc environments. Their approach effectively tackles both process and link failures through a unified mechanism within the Heard-of (HO) model, particularly suited for environments with transient faults. This methodology is crucial as transient link faults, including message losses, and process faults, such as temporary unavailability due to signal obstructions, are common in ad hoc networks.

To validate their theoretical advancements, Borran et al. conducted comprehensive simulations using the JiST/SWANS wireless network simulator. These simulations spanned a variety of scenarios, including both single and multi-hop networks, to assess the performance and resilience of their enhanced consensus algorithm. The results reveal that their modified Paxos/LastVoting algorithm, integrated with the new communication layer, achieves consensus effectively even in environments with significant message loss and node mobility. These findings substantiate the practical applicability of their strategy and provide insights into the behavior of consensus algorithms in realistic wireless settings.

III. Leader Election Algorithm

Leader election algorithms, as a special kind of consensus algorithm, also had considerable theoretical research in the early stages. These studies played an important role in promoting the development of wireless consensus algorithms. A lock-step leader election mechanism was developed in the research on the Secure Extrema Finding Algorithm (SEFA). This algorithm is capable of evolving into a secure preference-based leader election model through the use of a utility function that captures the preferences of nodes within wireless ad hoc networks. The method integrates sophisticated cryptographic protocols to ensure robust security and confidentiality of communications during the election process [28]. A fully distributed leader election scheme was proposed to ensure that election values in wireless sensor networks are unforgeable and resistant to jamming attacks. Employing efficient symmetric key operations, the scheme guarantees that well-connected participating cluster members unanimously elect the same leader. Moreover, it secures the leader election process against potential manipulation by attackers who might seek to alter the outcome by disproportionately affecting the election chances of benign nodes. Additionally, the protocol is designed to rapidly recover from interruptions such as message losses or deliberate attacks, thereby enhancing the reliability and resilience of the electoral process in challenging environments [29]. A top k-leader election algorithm with faulty nodes was presented in [30], specifically designed to effectively manage faulty nodes within the system. This algorithm ingeniously ensures that the leader election process remains reliable even in the presence of node failures, by incorporating robust mechanisms to mitigate the impact of these faults. It enhances the resilience of the system by maintaining the integrity of the election process in dynamic and potentially unstable network environments, where nodes are susceptible to sporadic failures and disconnections. Gołębiewski and Klonowski [31] introduced an advanced leader election mechanism for ad hoc single-hop radio sensor networks, engineered to withstand adversarial actions such as continuous transmissions aimed at blocking channels and attempts to compromise the system by forging identities. Richa et al. [32] addressed the challenges of jamming attacks in single-hop wireless networks by focusing on the self-stabilization of the leader election process, which was achieved through dynamically adjusting the nodes' transmission probabilities at the MAC layer to enhance both robustness and reliability.

1.1.2.2 Practical Level

Later, researchers applied consensus algorithms to real-world systems, focusing on improving their performance in practical applications. Among the consensus algorithms deployed in actual systems, synchronous transmission-based consensus algorithms hold significant importance. Additionally, many researchers are exploring ways to enhance consensus algorithm performance by integrating trusted execution environment (TEE), randomness, random timers, Proof-of-Channel, and Spanner Tree. Reducing energy consumption and efficiently utilizing memory are also major research hotspots.

I. Synchronous Transmission
Traditional low-power wireless protocols have considered avoiding packet collisions as a cornerstone of their design, but the concept of synchronous transmission challenges this viewpoint. Under certain conditions, generic low-power radios can receive useful information even with overlapping signals from different transmitters. Synchronous transmission achieves effective information transfer through the capture effect and constructive interference. This approach is particularly important in the Internet of Things (IoT) and Cyber-Physical Systems (CPS), as these systems often require efficient communication in environments with limited energy and bandwidth.

Capture effect is a phenomenon in wireless communications where a receiver can successfully decode the strongest signal amid multiple concurrent signals. This requires the dominant signal's strength to exceed the combined intensity of other signals and background noise by a specific threshold. A key advantage of the capture effect is its ability to enhance the probability of packet reception in environments with interference or multipath effects. Wireless network protocol designers leverage this attribute to improve the robustness and reliability of data transmission in complex settings.

Constructive interference occurs when multiple wireless signals arrive at a receiver almost simultaneously and are precisely aligned in phase, allowing them to cumulatively form a stronger signal. Key to successfully harnessing constructive interference is ensuring exact synchronization of the transmission timing and phase alignment. This typically requires highly precise clock synchronization and coordinated transmission mechanisms. Constructive interference enables receivers to enhance signal quality with the aid of multiple transmitting sources, especially in environments with weak signals or high packet loss rates.

Wireless 2PC & Wireless 3PC In the realm of low-power wireless networks, achieving network-wide consensus is essential for applications such as cooperative robots and industrial control systems. Al Nahas et al. [33] introduced "A^2: Agreement in the Air," a system engineered to enable distributed consensus in multi-hop networks via a novel synchronous transmission kernel named Synchrotron. This kernel leverages the capture effect, frequency hopping, and link-layer security to establish a robust communication infrastructure. Their key innovation lies in integrating two-phase and three-phase commit protocols to enhance both the reliability and speed of consensus decisions, augmented by mechanisms for group membership management, sequence distribution, and cryptographic rekeying.

Al Nahas et al. tackle the challenge of adapting traditional wired consensus protocols for low-power wireless environments, a topic not extensively explored in prior research. By utilizing synchronous transmission technologies like Glossy and Chaos, they significantly decrease the latency associated with achieving consensus wirelessly, thereby facilitating real-time applications that demand rapid and dependable agreement mechanisms. Moreover, the adoption of frequency hopping and the use of multiple parallel channels increase the network's resilience to interference and failures.

The team rigorously tested the effectiveness of A^2 across four public testbeds with varying node densities and sizes. Their experiments confirmed that A^2 could deliver consensus with exceptional speed and reliability, executing a two-phase commit in just 475 milliseconds across 180 nodes while maintaining a duty cycle as low as 0.5% during one-minute intervals. These results underscore A^2's ability to ensure zero data loss end to end, even in prolonged tests, which underscores its potential for mission-critical applications under demanding network conditions.

In summary, the A^2 system presents a reliable and efficient method for achieving network-wide consensus in low-power wireless networks. Through their innovative application of synchronous transmissions and robust network services, Al Nahas et al. address substantial challenges related to latency and reliability, essential for deploying next-generation cooperative and control applications in wireless environments.

Wireless Paxos Poirot et al. [34] tackle the issue of achieving network-wide consensus in multi-hop low-power wireless networks, environments inherently constrained by limited resources and inefficiencies of traditional methods. They introduce "Wireless Paxos," a creative reimagining of the traditional Paxos consensus protocol, specifically revamped to perform effectively within the confines of low-power wireless networks. By transforming the traditionally unicast-based communications of Paxos into a dynamic many-to-many communication scheme, they effectively harness the intrinsic broadcast capabilities of wireless networks. This redesign substantially decreases the time required to reach consensus and simultaneously boosts the protocol's reliability and fault tolerance, even amid node or communication failures.

In rigorous testbed experiments involving up to 188 nodes, they demonstrate that their protocol can reach consensus in just 289 milliseconds–far surpassing the speed of conventional methods. Moreover, the protocol's durability is put to the test with controlled failures, showcasing its ability to maintain consistency and function optimally under challenging conditions. Extensive evaluations on physical deployments further assess Wireless Paxos's performance, revealing significant enhancements in latency and reliability when compared to existing solutions. These findings confirm the protocol's efficacy and efficiency, establishing it as a robust consensus mechanism for low-power wireless networks.

Wireless PBFT Goyal et al. [35] tackle the challenge of enabling Byzantine fault tolerance in resource-constrained IoT systems by introducing the ReLI framework. They design a real-time, lightweight consensus mechanism tailored for energy-efficient operations in low-power IoT environments. This innovative framework incorporates synchronous transmission techniques to significantly enhance communication efficiency and reduce energy costs, which is crucial for the sustainability and reliability of battery-operated IoT devices.

Their empirical validations showcase notable performance enhancements with ReLI over traditional Byzantine fault tolerance solutions. Through comprehensive testing on a recognized IoT/WSN testbed, they confirm that their framework operates significantly faster and consumes less energy. These tests validate ReLI's ability to achieve up to 80% faster operational speeds and up to 78% reductions in radio-on time, underscoring its improved efficiency and performance.

Moreover, the team substantiates the practicality of ReLI by not only demonstrating superior performance but also detailing the technical adaptations needed to effectively integrate synchronous transmission mechanisms. These modifications enable the framework to manage the complexities associated with Byzantine faults more efficiently and with less energy consumption. Their approach provides a significant contribution to the fields of distributed computing and IoT by offering a scalable and robust solution for achieving consensus in decentralized systems vulnerable to malicious threats.

II. Trusted Execution Environment (TEE)

TEE is a secure area within a main processor designed to ensure the integrity and confidentiality of data and code. TEEs provide an isolated environment that runs parallel to the main operating system, allowing sensitive computations to be executed securely, even if the main system software is compromised. ARM TrustZone and Intel SGX are popular TEE implementations, offering robust frameworks to protect sensitive operations from malicious attacks and unintentional interference. TEEs play a crucial role in various applications, including secure payment systems and digital rights management, as highlighted in recent research, enhancing fault tolerance in distributed systems by preventing equivocation attacks and ensuring data authenticity.

Xu et al. [36] address the challenge of Byzantine fault tolerance in wireless embedded systems through the RATCHETA algorithm. They enhance fault resilience and reduce message complexity by using a hybrid fault model incorporating a small trusted subsystem based on ARM TrustZone. This subsystem hosts a pair of monotonic counters to prevent equivocation attacks, allowing the system to tolerate up to $\frac{n-1}{2}$ Byzantine faults. Their innovation lies in the dual-counter authentication mechanism, which ensures bounded memory usage and message size. They implemented RATCHETA on Raspberry Pi devices in an ad hoc wireless network to validate its performance. The experimental results demonstrated robustness against Byzantine attacks and efficient consensus achievement under conditions of message loss and network unreliability, showcasing significant advancements in fault tolerance for small-scale autonomous systems.

Xu et al. [37] present a framework for Byzantine fault-tolerant consensus in small, wirelessly connected autonomous systems. They focus on achieving consensus in scenarios such as vehicle platooning and UAV swarms, where environmental conditions are harsh and safety is critical. Their framework supports both general-purpose value agreement and distributed sensor readings by utilizing a trusted subsystem that only fails by crashing, reducing the required participants from $3t + 1$ to $2t + 1$. This approach reduces message complexity and improves efficiency. Experimental validation on Raspberry Pi devices in ad hoc wireless mode demonstrated the framework's robustness and efficiency, highlighting a significant improvement in the practicality of Byzantine fault-tolerant protocols for small-scale autonomous systems.

III. Random Mechanisms

When considering random mechanisms used in consensus algorithms, we can mainly divide them into two categories. The first category achieves consensus in an asynchronous network environment by introducing randomness. The second category uses a random timer to improve the efficiency of consensus. These two types of random mechanisms demonstrate their respective advantages in different network environments and application scenarios, becoming important consensus tools in distributed systems.

Randomness In the context of wireless consensus algorithms, randomization techniques have emerged as a critical response to the challenges of achieving reliable agreement in asynchronous and failure-prone environments. Traditional deterministic algorithms often fall short due to the well-known FLP impossibility result, which states that reaching consensus in an asynchronous system is impossible if even a single node can fail. This challenge is further compounded in wireless ad hoc networks, where nodes not only face arbitrary failures (Byzantine faults) but also encounter dynamic communication disruptions such as message collisions, interference, and disconnections caused by mobility.

Randomization provides a powerful solution by introducing probabilistic elements into the decision-making processes of consensus algorithms. By using local random bit generators, each node makes independent, unbiased decisions that help break symmetry and facilitate convergence even under adverse conditions. This approach ensures that while consensus may not be reached in a fixed number of steps, it will eventually be achieved with probability one. Randomization mitigates the effects of dynamic message losses and Byzantine behaviors without relying on costly cryptographic operations, thus making the algorithms more suitable for resource-constrained and unreliable wireless environments. This innovative technique significantly enhances the resilience and efficiency of consensus protocols, rendering them robust against the inherent uncertainties of wireless ad hoc networks.

Moniz et al. [38] investigated the problem of achieving Byzantine consensus in resource-constrained wireless ad hoc networks. They developed the Turquois protocol, specifically designed to tolerate Byzantine faults and dynamic message omissions without relying on synchrony assumptions or public-key cryptography during normal operations. This protocol allows nodes to reach consensus even in the presence of arbitrary failures, leveraging the natural broadcasting medium to enhance efficiency. By avoiding traditional point-to-point communication models and embracing the unreliable nature of wireless communications, they created a more robust solution for consensus in these challenging environments.

The innovation lies in the use of randomization to ensure progress despite the inherent unreliability and potentially malicious behavior in wireless ad hoc networks. Each node uses a local random bit generator to make independent decisions, helping to break symmetry and facilitate convergence toward consensus. The Turquois protocol guarantees safety and liveness, ensuring that consensus is eventually reached with probability one. Validation of the protocol was achieved through extensive experimental evaluation, comparing its performance against two well-known intrusion-tolerant consensus protocols in various network and fault scenarios. The results demonstrated that Turquois significantly outperforms these protocols, particularly as the system scales, proving the effectiveness and efficiency of their approach.

Random Timer Random timer technology, also known as the backoff mechanism, is a fundamental component in wireless communication protocols such as IEEE 802.11. This technology emerged to address the challenge of data packet collisions in shared communication channels. When multiple nodes attempt to transmit data simultaneously, collisions occur, leading to data loss and the need for retransmissions, which significantly degrades network performance.

To mitigate this issue, the IEEE 802.11 protocol implemented a backoff mechanism where each node waits for a random period before attempting to transmit. This waiting period is determined by a random timer set within a contention window that dynamically adjusts based on network conditions. By randomizing the waiting time for each node, the probability of simultaneous transmissions is reduced, thereby minimizing collisions. This mechanism enhances overall network efficiency by ensuring fair access to the communication medium and reducing the frequency of collisions. The random timer approach is highly effective in handling data packet collisions and adapts well to varying network traffic conditions, making it an essential technique in modern wireless communications.

Zhou et al. [39] investigate the performance of wireless practical Byzantine fault tolerance (PBFT) networks utilizing the IEEE 802.11 protocol. They address the challenge of implementing PBFT consensus mechanisms in wireless networks, where existing research has predominantly focused on wired scenarios. The primary objective is to evaluate how wireless communication impacts the success probability, delay, and throughput of PBFT networks under unsaturated traffic conditions. Their work aims to bridge the gap in understanding the performance of PBFT in wireless environments, especially considering the inherent instability and limited spectrum resources of such networks.

They introduce an innovative application of the IEEE 802.11 protocol's broadcast scheme to PBFT, avoiding the use of acknowledgment (ACK) and request-to-send/clear-to-send (RTS/CTS) messages, which are typically employed to enhance transmission reliability. This novel approach leverages the physical sensing capabilities of IEEE 802.11 to manage communication without the overhead of these additional messages. They develop a comprehensive performance model based on a Markov chain to analyze the protocol's success probability, transaction confirmation delay, and throughput. By simulating various network configurations and traffic conditions, they validate their model and demonstrate its accuracy.

To verify their theoretical analysis, Zhou et al. conduct extensive simulations illustrating the relationships between the number of nodes, packet arrival rates, contention window sizes, and performance metrics. Their results show that while the wireless PBFT network is resilient to a certain degree of collision and packet loss, its performance significantly declines as the network scale increases beyond a certain threshold. This study provides valuable insights and practical guidance for deploying PBFT-based wireless blockchain networks using the IEEE 802.11 protocol.

IV. Proof-of-Channel

PoC technology emerged as a solution to the limitations of traditional blockchain consensus mechanisms, such as Proof-of-Work (PoW) and Proof-of-Stake (PoS), in wireless networks. In scenarios where nodes have limited physical resources and face variable bandwidth due to environmental factors or adversarial attacks, conventional consensus algorithms often fall short. PoC leverages the inherent properties of wireless communications, specifically channel competition, to establish consensus. This innovative approach seamlessly integrates the process of channel competition with the consensus mechanism, effectively reducing communication overhead and enhancing consensus efficiency. By having nodes compete for channel access to propose new blocks, PoC ensures that the consensus process is inherently tied to the current state of the wireless environment, making it more robust against issues like jamming and interference. This method significantly improves the security and reliability of blockchain protocols in wireless networks, addressing the challenges posed by the lack of stable, reliable channels and the susceptibility to jamming attacks.

Xu et al. [40] address the challenges of implementing blockchain protocols in single-hop wireless networks under adversarial conditions. Traditional consensus mechanisms like Proof-of-Work (PoW) and Proof-of-Stake (PoS) are ill-suited for wireless environments where nodes face limited resources and variable bandwidth due to interference and jamming. To overcome these issues, they propose a novel consensus protocol called Proof-of-Channel (PoC), which leverages the inherent properties of wireless communications, specifically the competition for channel access. By integrating channel competition into the consensus process, PoC reduces communication overhead and enhances the efficiency and robustness of the consensus mechanism in adversarial environments.

The authors developed a permissioned blockchain protocol, BLOWN (BLOckchain protocol for Wireless Networks), based on PoC, specifically designed for single-hop wireless networks operating under a signal-to-interference-plus-noise-ratio (SINR) model. They formalize BLOWN using the universal composition framework and rigorously prove its security properties, such as persistence and liveness. Additionally, BLOWN demonstrates resilience against adversarial jamming, double-spending, and sybil attacks. To validate their approach, Xu et al. conducted extensive simulation studies. These simulations show that BLOWN effectively mitigates the challenges posed by the wireless environment, ensuring secure and reliable blockchain operations. The innovative PoC mechanism and comprehensive validation through simulations highlight the practicality and effectiveness of BLOWN in enhancing the applicability of blockchain technology in wireless networks.

V. Spanner Tree

Spanner trees, serving as the communication backbone in wireless consensus protocols, have emerged to tackle the unique challenges posed by multi-hop wireless networks. Traditional consensus algorithms, originally designed for wired networks, often struggle with the high message complexity and interference issues inherent in wireless environments. To mitigate these challenges, spanner trees offer a sparse yet efficient network structure that balances simplicity and communication overhead. By organizing nodes into hierarchical levels, spanner trees reduce the number of necessary links, thus minimizing interference and enhancing data aggregation efficiency. Each level of the spanner tree aggregates data from the previous level, ultimately funneling information to a central node or root. This hierarchical approach ensures that data can be efficiently collected and disseminated across the network, significantly improving the performance of consensus protocols in wireless settings. The implementation of spanner trees effectively addresses the problems of medium contention and high message complexity, providing a robust framework for achieving reliable consensus in wireless networks.

Xu et al. [41] address the challenges of designing efficient and fault-tolerant blockchain protocols for multi-hop wireless networks. Their work tackles the issues of medium contention and high message complexity inherent in wireless environments. They propose the wChain protocol, which integrates wireless communication properties with blockchain technologies under a realistic SINR model. By utilizing a hierarchical spanner as the communication backbone, they achieve fast data aggregation within $O(\log N \log \Gamma)$ slots, where N is the network size and Γ is the ratio of the maximum to minimum distances between nodes. Additionally, wChain employs mechanisms for data aggregation, reaggregation, and node recovery to ensure efficiency, fault tolerance, persistence, and liveness.

The innovation of wChain lies in its ability to handle fault tolerance through an optimal spanner construction that allows efficient communication and data aggregation. The worst-case runtime of wChain is upper bounded by $O(f \log N \log \Gamma)$, where f is the maximum number of faulty nodes. Xu et al. validate their design through both theoretical analysis and extensive simulation studies. The theoretical analysis confirms the protocol's properties, while the simulation results demonstrate wChain's superior performance in terms of efficiency and fault tolerance. These experiments are conducted under various network conditions and parameters, showcasing wChain's robustness and applicability to real-world scenarios in multi-hop wireless networks.

VI. Resource Optimization

Reducing the resources required for consensus algorithms has significant advantages, including lowering operational and maintenance costs, improving system scalability, promoting green computing and sustainability, and better adapting to low-power scenarios. These optimizations can significantly enhance the overall performance and practicality of consensus algorithms, allowing them to be widely applied in more fields.

Energy The development of energy-efficient consensus protocols for wireless networks addresses the crucial need to minimize energy consumption while maintaining reliable communication and fault tolerance. Traditional Byzantine fault-tolerant state machine replication (BFT-SMR) protocols often focus on reducing communication complexity, improving throughput, or lowering latency. However, these protocols typically incur significant energy costs due to the extensive use of cryptographic operations and high communication overhead. In wireless ad hoc networks, where nodes are often battery powered, optimizing energy usage is essential to prolong the system's operational lifetime and ensure sustainability.

Energy-efficient BFT-SMR protocols meet this challenge by optimizing the steady state, where the leader node is assumed to be correct. This optimization is achieved by reducing the number of required cryptographic signatures and minimizing communication complexity, significantly lowering energy consumption. Additionally, leveraging multicast communication, where a single transmission can reach multiple nodes, further enhances energy efficiency. By shifting computationally expensive operations to less frequent phases, such as the view-change phase, these protocols balance the trade-off between energy efficiency and fault tolerance, ensuring reliable performance in resource-constrained environments.

Bhat et al. [42] tackle the critical issue of energy efficiency in BFT-SMR protocols, focusing on wireless ad hoc networks where nodes are often battery powered. They introduce a novel protocol called EESMR, which optimizes energy usage in the steady state when the leader node is correct. This optimization is achieved by reducing the number of required cryptographic signatures and minimizing communication complexity. Additionally, EESMR leverages multicast communication, where a single transmission can reach multiple nodes, further enhancing energy efficiency. By shifting computationally expensive operations to less frequent phases, such as the view-change phase, EESMR maintains a balance between energy efficiency and fault tolerance.

The innovative aspect of their work lies in the combination of reducing cryptographic overhead and utilizing multicast communication to significantly lower energy consumption. Bhat et al. developed a comprehensive energy model to analyze and compare the energy efficiency of their proposed protocol against existing state-of-the-art solutions. Their findings demonstrate that EESMR can achieve up to 64% energy savings in certain scenarios.

To validate their approach, Bhat et al. implemented EESMR on a test bed of wireless nodes using Bluetooth low energy (BLE) modules. Through extensive experiments, they measured the energy consumption of their protocol and compared it with Sync HotStuff, a leading BFT-SMR protocol. The results confirmed their theoretical analysis, showing that EESMR is substantially more energy-efficient during failure-free runs. This practical demonstration underscores the feasibility and benefits of their proposed energy-efficient consensus protocol for resource-constrained environments.

Memory In the domain of wireless consensus protocols, particularly within embedded systems, the limitation of available memory resources presents a substantial challenge. Traditional Byzantine fault-tolerant (BFT) protocols, primarily designed for server-grade hardware, necessitate considerable memory and computational capacities that far exceed what is available on devices constrained by memory, such as those prevalent in Internet of Things (IoT) deployments. Typically, these systems operate with less than 1 MB of RAM–a stark contrast to the vast resources equipped on conventional servers. To address the critical need for memory optimization in such constrained environments, researchers have innovated methods to minimize the memory demands of BFT protocols. These advancements include the adoption of static memory allocation techniques, allowing for a predetermined maximum memory usage at compile time, and the development of memory-efficient consensus algorithms tailored to the stringent resource limitations of microcontrollers and other similar devices.

Böhm et al. [43] have identified a method to solve the significant challenge of implementing BFT consensus protocol in environments that are severely limited by memory resources. They present TINYBFT, an innovative BFT state-machine replication library specifically developed for devices constrained to a maximum of 1 MB of RAM. A key innovation in their work is the memory-efficient adaptation of the PBFT protocol, which uniquely employs static memory allocation to ensure that the upper bounds of memory usage are established at compile time. This method markedly contrasts with the conventional approach of BFT protocols, which are typically engineered for server-grade hardware equipped with substantially larger memory capacities.

Their research focuses on adapting BFT protocols to operate efficiently on microcontrollers, such as the ESP32-C3, commonly found in embedded systems for Internet of Things (IoT) applications. By redesigning the PBFT protocol to reduce memory demands without sacrificing performance, Böhm et al. guarantee that their protocol can function within the strict memory constraints of the targeted devices. They conducted thorough experiments to validate the functionality of TINYBFT, showcasing its capability to manage consensus operations with minimal latency on these resource-restricted platforms. The empirical evaluation involved deploying TINYBFT on actual ESP32-C3 microcontrollers and monitoring its performance across various operational scenarios. The findings underscored that TINYBMBFT could sustain reliable consensus mechanisms without breaching the memory and processing limits of the microcontrollers. Through their experiments, Böhm et al. not only demonstrated the feasibility of their approach but also established a foundational benchmark for future endeavors in integrating BFT protocols into resource-limited embedded systems.

1.1.3 Simulators and Testbeds for Evaluation

The experiment environment for wireless consensus algorithms can be categorized into two types: deployment on simulators and deployment on practical devices. Simulators offer advantages such as low cost, high flexibility, strong security, rapid iteration, and high reproducibility, making it easy to test and adjust algorithms under various network conditions and device configurations. In contrast, deployment on actual devices provides more realistic test results, helps identify issues not foreseeable by simulators, provides accurate performance evaluation data, and verifies the algorithm's adaptability in real-world environments. Therefore, by conducting tests in both types of environments, one can comprehensively evaluate the performance and reliability of wireless consensus algorithms, thereby optimizing their design and enhancing their performance in practical applications.

1.1.3.1 Simulator

Simulators have advantages in simulating consensus algorithms, such as ease of operation, high flexibility, low cost, and the ability to easily simulate various network environments. These benefits allow researchers to quickly verify and adjust algorithm performance. Additionally, these tools often come with detailed documentation and active community support, further lowering the barrier to use. However, simulators also have some limitations, such as the inability to fully replicate real-world environments, a lack of real network randomness and interference factors, and potential difficulties in accurately simulating the complex behavior of large-scale networks in some cases.

OMNeT++[1] is a network simulator developed in C++, engineered to facilitate the construction and simulation of complex communication networks. It features

[1] https://github.com/omnetpp

robust visualization and comprehensive network analysis tools, making it exceptionally well suited for simulating environments ranging from wireless and wired networks to distributed systems.

NS3,[2] an open-source network simulation tool, facilitates simulations spanning from the physical layer to the application layer. It is particularly advantageous for the study of wireless communication protocols. NS3 offers programming interfaces in both C++ and Python, enabling the simulation of various wireless technologies, including LTE, Wi-Fi, and WiMAX.

Emulab[3] functions as a network testbed that integrates simulation with physical testing, thereby permitting experiments within a controlled setting. It is adept at handling complex network research and testing scenarios and is capable of simulating expansive networks, both wired and wireless.

1.1.3.2 Practical Testbeds

Compared to simulators, deploying wireless consensus algorithms in real-world environments has significant advantages, including:

- **Realistic Node and Network Behavior:** Node and network behavior are driven by actual hardware and environmental conditions, providing genuine data and performance.
- **Random Factors and Uncertainty:** The presence of wireless signal interference, node failures, and physical obstacles introduces random factors and uncertainty, reflecting the algorithm's performance in complex dynamic environments.
- **Accurate Performance Evaluation:** Real-world deployment allows for the accurate assessment of critical metrics such as latency, throughput, reliability, and energy consumption.
- **Hardware Compatibility and Performance Testing:** Testing on different hardware platforms ensures the consensus algorithm's compatibility and performance.
- **Adaptability to Environmental Changes:** It evaluates the algorithm's adaptability to environmental changes such as temperature, humidity, and geographic conditions, offering essential feedback on its robustness.

FlockLab Developed by ETH Zurich, it serves as a dedicated testbed for researching and validating the hardware, software, and algorithms associated with wireless sensor networks and embedded systems. The platform provides an array of features including real-time monitoring, energy consumption analysis, and precise localization capabilities. Accessible remotely, FlockLab allows researchers to conduct extensive and complex experiments, particularly focusing on scenarios that involve

[2] https://www.nsnam.org/

[3] https://www.emulab.net/portal/frontpage.php

interactions across multiple nodes, thereby facilitating advanced studies in network behavior and performance.

FIT-IoT-LAB Developed collaboratively by several French research institutions, including INRIA and Grenoble INP, FIT-IoT-LAB is an open testbed designed to advance the research and testing of IoT devices and their applications. The lab boasts over 2000 wireless nodes spread across various locations, each equipped with diverse processing powers and wireless technologies. It supports a wide range of experiments from fundamental performance evaluations to intricate system-level integrations. FIT-IoT-LAB is renowned for its user-friendly interface and broad support for numerous programming languages, making it an invaluable resource for researchers looking to deploy and assess their IoT solutions efficiently.

Table 1.4 presents some of the devices used in existing research work. These devices play critical roles in various studies, assisting researchers in testing and validating the performance and effectiveness of different wireless consensus algorithms. By analyzing these devices in detail, we can understand their applicability and advantages in wireless consensus algorithms, providing valuable reference and guidance for future research.

- The Raspberry Pi is a low-cost, credit-card-sized single-board computer developed by the Raspberry Pi Foundation in the United Kingdom, widely used in education, research, and hobbyist electronic projects. It is equipped with various interfaces and can run multiple operating systems, making it suitable for a variety of programming and electronic experiments. The Raspberry Pi 3 Model B [36] is equipped with a quad-core ARM Cortex-A53 1.2 GHz processor, 1 GB RAM, and a 2.4 GHz 802.11n wireless module, offering robust performance suitable for applications requiring higher processing power and wireless communication capabilities.
- The NUCLEO development board is a tool based on STMicroelectronics microcontrollers, featuring an integrated debugger, various interfaces, and extensive software support, making it suitable for a wide range of embedded applications, particularly excelling in energy efficiency and wireless communication. The NUCLEO-F401RE [42] model is equipped with an STM32F401RE microcontroller, 512 kB of flash memory, 96 kB of SRAM, and a low-power Bluetooth module, utilized for wireless communication and energy efficiency testing.

Table 1.4 Comparison of experimental platforms

Platform	CPU	Memory	Communication technology
Raspberry Pi 3 (Model B)	Quad-core ARM cortex-A53 1.2 GHz	1 GB RAM	2.4 GHz 802.11n
NUCLEO-F401RE	ARM cortex M4 84 MHz	512 kB flash /96 kB SRAM	Bluetooth low energy (BLE)
ESP32C3	160 MHz	Not specified	2.4 GHz Wi-Fi 802.11 b/g/n

- The ESP32 is a high-performance, low-power microcontroller featuring a dual-core Xtensa LX6 processor with a maximum frequency of 240 MHz, integrating 2.4 GHz Wi-Fi and Bluetooth 4.2/BLE. It supports various peripheral interfaces such as SPI, I2C, UART, ADC, and PWM, and is widely utilized in Internet of Things (IoT) applications. The ESP32-C3 [43], a variant within this series, incorporates a single-core RISC-V processor operating at 160 MHz, supporting 2.4 GHz Wi-Fi and Bluetooth 5.0, making it suitable for resource-constrained embedded systems.

1.2 Wired Versus Wireless Consensus Protocols

We compare wired consensus protocols and wireless consensus protocols from three perspectives: physical layer, network layer, and consensus layer. The physical layer and network layer are used to compare the differences between wireless and wired networks. We refer to the Transmission Control Protocol/Internet Protocol (TCP/IP) five-layer network model. The physical layer corresponds to the physical layer in the five-layer model. Our comparison for the network layer encompasses the four layers above the physical layer in the TCP/IP five-layer model. For the consensus layer, we compare the system models, impossibility results, and research directions of the two types of consensus protocols. The TCP/IP five-layer model is as follows:

1. **Physical Layer:** Responsible for the transmission of raw data through physical mediums like cables and radio waves.
2. **Data Link Layer:** Ensures reliable data transfer between directly connected nodes by providing error detection and correction mechanisms.
3. **Network Layer:** Manages packet routing across multiple networks using logical addressing and ensures the correct pathfinding for data.
4. **Transport Layer:** Provides end-to-end communication services and manages data packet segmentation and reassembly to ensure data integrity.
5. **Application Layer:** Facilitates user interaction with network services, handling specifics of application operations to meet user needs.

1.2.1 Physical Layer

As shown in Table 1.5, wired networks and wireless networks each have their unique advantages and limitations. Wired networks, such as Ethernet, typically use copper or fiber optic cables for data transmission. A major advantage of this type of network is its high speed and low latency, which are crucial for data-intensive tasks and applications that require quick responses. Additionally, wired connections usually offer higher security, as physical access points are easier to control, reducing the risk of data interception. Stability is also a key advantage of wired networks, as they are

Table 1.5 Comparison of wired and wireless networks

	Wired	Wireless
Communication medium	Copper, fiber, etc.	Radio waves
Mobility and roaming	Limited	Higher
Security	High	Lower than wired. Easier to hack
Reliability	High	Lower than wired
Inference	Low	High
Cost	High initial cost, low maintenance cost	Low initial cost, high maintenance cost
Benefits	*High data rate	*Cable-free convenience
	*Strong noise resilience	*Suitable for mobile devices
	*Highly reliable	*Easy installation
	*Greater security	*Flexibility and scalability

less susceptible to environmental interference, such as interference from wireless signals or weak coverage. However, the main limitation of wired networks is their poor mobility—users need to be connected to physical cables to access the network. Furthermore, the installation of wired networks typically requires more time and cost, especially when deploying a large number of network cables inside buildings.

In contrast, wireless networks, such as Wi-Fi or cellular networks, transmit data through the air, providing greater flexibility and mobility. They allow users to move freely within the coverage area without the constraints of physical connections. This makes wireless networks highly suitable for mobile devices and temporary installation scenarios. Installing and maintaining wireless networks is usually simpler and more cost-effective than wired networks, as they do not require complex cable layouts. However, the disadvantage of wireless networks is that they usually offer lower data transmission speeds, especially in environments with weak signals or a lot of interference. The security of wireless networks is also generally not as good as that of wired networks, as wireless signals can be intercepted by any device within the coverage range. Additionally, wireless networks are more susceptible to interference from physical obstacles (such as walls) and other wireless devices.

1.2.2 Network Layer

The comparison of the network layer is divided into three parts: basic communication protocol, advanced communication protocol, and network architecture. In terms of basic communication protocol, the comparison includes communication technologies and fundamental communication primitives (such as routing, multicasting, and broadcasting). These basic communication protocols in consensus algorithms provide a stable fundamental communication mechanism, ensuring efficient and reliable information exchange between nodes, thereby supporting the realization of consensus. Advanced communication protocol focuses on the comparison of resource management and allocation, and congestion control. The role of advanced communication protocols in consensus is to optimize the use of network resources, ensure effective communication under high load conditions, reduce delay and packet loss, and thus improve the efficiency and reliability of consensus algorithms. In terms of network architecture, the comparison includes the organizational structure of network nodes. Different network architectures can affect the design and implementation of consensus algorithms. The role of network architecture in consensus is to determine how nodes are connected and interact, thereby affecting the propagation speed of consensus, fault tolerance, and overall performance.

1.2.2.1 Basic Communication Protocol

The following introduces the basic communication protocols of wired networks. For the basic communication protocols of wireless networks, please refer to Sect. 1.1.1.1.

I. Communication Technologies

Ethernet is a widely used network technology that primarily encompasses protocol standards for the physical and data link layers. At the physical layer, Ethernet specifies various aspects including cable types such as coaxial, twisted pair, and fiber optics, as well as signal transmission and hardware interface standards. Within the data link layer, Ethernet employs MAC addresses to identify devices and define frame structure, encapsulation, effective frame transmission, and error detection mechanisms. The Ethernet standard does not directly address protocols pertaining to the network, transport, and application layers. In practice, protocols that operate at these levels, such as TCP/IP, are handled by higher-level protocols, with Ethernet providing the foundational data transmission services.

IEEE 802.3 is a technical specification that defines Ethernet communication standards. With the proliferation of Ethernet technology, the IEEE adopted and standardized this technology, resulting in the IEEE 802.3 standard. This standard expands and formalizes the original Ethernet technology, specifying a variety of technical specifications for the physical and data link layers, including cable types, speed ratings, and transmission ranges. IEEE 802.3 progressively incorporates support for new technologies, such as full-duplex mode and high-speed Ethernet, including Gigabit and 10 Gigabit Ethernet. IEEE 802.3 can be regarded as the standardized version of Ethernet technology, not only incorporating the basic framework of the original Ethernet but also evolving and expanding to meet the changing demands of network requirements.

Remote direct memory access (RDMA) optimizes data center performance by enabling network devices to access another computer's memory directly, bypassing the CPU. Primarily, RDMA technologies engage at the physical, data link, and transport layers, with limited interaction at the network and application layers. Below are succinct descriptions of three principal RDMA implementations: InfiniBand, RoCE, and iWARP.

- **InfiniBand:** Employed predominantly in data centers and high-performance computing, InfiniBand is a high-performance network architecture. It operates with specialized cables and switches at the physical layer and utilizes proprietary transmission protocols at the data link layer to ensure reliable data transfer. At the transport layer, InfiniBand provides sophisticated control and error-handling mechanisms, though it delegates the handling of the network and application layers to upper-level IP protocols and application processes.
- **RoCE (RDMA over Converged Ethernet):** RoCE enables RDMA functionalities over standard Ethernet infrastructures, integrating seamlessly with Ethernet technology at both the physical and data link layers. Distinct from InfiniBand, RoCE involves network layer operations such as IP routing and congestion control, which introduce a layer of complexity in standard Ethernet environments but also enhance data transmission efficiency at the transport layer.
- **iWARP (Internet Wide Area RDMA Protocol):** iWARP facilitates RDMA over conventional TCP/IP networks, thus broadening the scope of remote direct memory access across extensive networks. It harnesses the existing network hardware at the physical and data link layers and adapts the TCP protocol at the transport layer to accommodate RDMA operations, thus capitalizing on existing network infrastructure without necessitating specialized hardware.

These technologies strive to diminish latency and boost data transfer efficiency by facilitating direct memory-to-memory transfers, effectively circumventing the operating system.

II. Basic Communication Primitives

Basic Communication Primitives are fundamental to implementing consensus algorithms. They provide the essential mechanisms for information exchange and coordination between nodes. Through these basic communication primitives, nodes can efficiently propagate information, synchronize states, and reach consensus. Therefore, the design and optimization of basic communication primitives are crucial for the successful implementation of consensus algorithms.

- Routing in networks is a mechanism for determining the optimal path for data packets from one network node to another. This process relies heavily on routing tables, which record the best paths to different target network addresses. Routing algorithms may consider various criteria such as path cost, hop count, and bandwidth to decide the best route. In complex networks, routing also manages dynamic path changes to ensure network connectivity and efficiency. Typical routing protocols include Open Shortest Path First (OSPF), Border Gateway Protocol (BGP), and Routing Information Protocol (RIP).
- Multicasting is a communication method that lies between unicast and broadcast, allowing information to be sent from one or multiple sources to a well-defined group of receivers. Unlike broadcasting, multicasting sends data only to those group members who have subscribed to a specific service, which significantly

conserves bandwidth and resources. Multicasting is particularly useful in applications such as video conferencing and real-time data transmission. Common multicasting protocols include Internet Group Management Protocol (IGMP), Protocol Independent Multicast (PIM), and Distance Vector Multicast Routing Protocol (DVMRP).

- Broadcasting is a communication mechanism that allows data or information to be sent from one node to all other nodes within the network. In wired networks like Ethernet, when a node sends a broadcast message, every device within the same broadcast domain, such as a subnet, receives this message. This method is commonly used for address resolution and service announcements but can lead to unnecessary network traffic and congestion. Ethernet and the Address Resolution Protocol (ARP) are typical protocols used for broadcast communications.

1.2.2.2 Advanced Communication Protocols

Advanced Communication Protocols cover two main aspects: resource management and allocation, and congestion control, both of which play a crucial role in enhancing network performance and reliability. Through effective resource management and congestion control, communication networks can maintain efficient operation in various complex and high-load environments. This ensures the stability of data transmission and the quality of service, thereby meeting the growing user demands and diverse application scenarios.

Resource Management and Allocation Resource management and allocation in wired networks is relatively straightforward due to fixed physical connections, which allow for precise control over bandwidth allocation and simplified congestion control through flow management mechanisms like TCP. The inherent stability of wired connections enables predictable and reliable network performance, thereby facilitating more effective resource management.

In contrast, in wireless networks, resource management and allocation are inherently complex and dynamic, necessitating the use of protocols such as CSMA/CA for collision avoidance and techniques like TDMA, FDMA, and CDMA for efficient channel allocation. To maintain reliability amid variability in bandwidth and throughput, advanced error detection and correction methods are also essential.

Congestion Control In wired networks, the TCP protocol exemplifies congestion control mechanisms, which incorporate algorithms such as slow start, congestion avoidance, fast retransmit, and fast recovery. These algorithms enable TCP to dynamically adjust the data transmission rate in response to network condition changes, thereby mitigating the risk of overload. Specifically, slow start and congestion avoidance carefully modulate bandwidth to prevent congestion, while fast retransmit and fast recovery swiftly address packet loss to reduce timeouts. Despite the enhancement of network reliability and throughput, TCP's performance may degrade during intense congestion or abrupt shifts in network conditions because it relies on end-to-end feedback, which can introduce delays in response.

Conversely, wireless networks deploy congestion control through channel-based and queue-based methods, supplemented by cross-layer design strategies. For instance, IEEE 802.11 protocols adapt transmission rates to match channel quality fluctuations, and LTE networks regulate congestion by adjusting resource allocation and scheduling priorities. These practices enable wireless networks to adapt their performance and resource utilization dynamically, tailored to the environmental conditions. However, congestion control in wireless environments is inherently more complex due to factors like the physical environment and device mobility, necessitating sophisticated coordination across multiple protocol layers. This not only increases the complexity and cost of implementation but also presents considerable challenges in maintaining network stability and responsiveness. In comparison to wired networks, wireless networks provide greater adaptability and flexibility in congestion management, yet they face significant challenges in maintaining consistent network performance and ensuring rapid response capabilities.

1.2.2.3 Network Architecture

The architectures of wired and wireless networks are distinctly different. Wired networks typically employ fixed topologies, connecting nodes through physical cables, which provide stable, high-bandwidth, and low-latency communication.

- **Bus Topology:** A bus topology connects all devices to a single central cable, known as the bus. This setup is straightforward and cost-effective to implement. However, it can be challenging to troubleshoot since any fault in the bus can disrupt the entire network, and performance may degrade with increased traffic.
- **Star Topology:** In a star topology, each device is linked to a central hub or switch. This configuration simplifies management and troubleshooting, as a failure in any single cable only affects the connected device. Nevertheless, the central hub constitutes a single point of failure, and the network's overall reliability depends on this node's functionality.
- **Ring Topology:** A ring topology connects each device to two other devices, forming a circular data path. This structure supports high-speed data transmission but is vulnerable to disruptions; a failure in any single device or connection can impact the entire network.
- **Tree Topology:** A tree topology combines elements of both star and bus topologies, with groups of star-configured networks linked to a linear bus backbone. This hierarchical structure is scalable and manageable but heavily relies on the backbone's integrity. Any compromise to the backbone can affect multiple subnetworks.

Wireless networks rely on radio waves for data transmission, resulting in more dynamic and flexible topologies where nodes can move freely. However, wireless networks generally have lower bandwidth, and higher latency, and are more susceptible to interference and signal attenuation.

- **Infrastructure Mode:** In infrastructure mode, wireless devices connect to a centralized access point (AP), which bridges them to a wired network. This mode offers centralized management and easy scalability. However, the AP serves as a single point of failure, and its range limits the network's coverage area.
- **Ad Hoc Mode:** Ad hoc mode allows wireless devices to communicate directly without a centralized access point. This flexible and easy-to-set-up configuration is ideal for temporary networks. Despite its simplicity, this mode struggles with scalability and management as the number of devices increases.
- **Mesh Topology:** Mesh topology involves each wireless node connecting to multiple other nodes, creating a Web-like network. This ensures high redundancy and fault tolerance, making the network self-healing. However, it is complex to configure and manage, with higher costs due to the extensive connections required.
- **Cellular Network:** Cellular networks consist of multiple base stations, each covering a geographic area known as a cell. These networks provide extensive coverage and support high mobility. Despite their scalability and wide reach, they are costly to establish and maintain, and they face challenges with limited frequency spectrum and potential interference.

1.2.3 Consensus Layer

We compare wired consensus and wireless consensus at the consensus layer from three aspects. First, there is the system model, which includes the network model and the adversary model. From the perspective of consensus algorithms, these models abstract the underlying network and adversaries, respectively. Second, there is the impossibility principle in consensus algorithms. The impossibility principle refers to conditions under which consensus algorithms cannot reach an agreement. Lastly, the common optimization techniques used by researchers in their work on these two types of consensus are considered.

1.2.3.1 System Models

Network Model Consensus algorithms in wired networks make relatively straightforward assumptions about the underlying network, primarily categorized into three types of synchrony: synchronous, partially synchronous, and asynchronous (refer to the Synchrony Abstraction in Sect. 1.1.1.2). Within these models, consensus algorithms in wired networks typically assume that any two nodes have a reliable authenticated point-to-point channel that does not drop messages. This assumption relies on stable communication mediums and advanced communication protocols, such as TCP/IP and Transport Layer Security (TLS). The TCP/IP protocol provides a stable and error-free message delivery mechanism, while the TLS protocol ensures communication security and message integrity. These assumptions significantly simplify the design of consensus algorithms, as developers can ignore many complexities in the underlying communication, focusing instead on the high-level logic and protocol optimization of the algorithms.

On the other hand, wireless consensus assumptions about the network model usually consider two aspects: network topology and the degree of interference. Network topology is divided into two aspects: whether the entire network is a single-hop or multi-hop network and whether the nodes are static or dynamic. The degree of interference is often measured using the signal-to-interference-plus-noise-ratio (SINR) model, which captures wireless network interference in a more realistic and precise way than a graph-based one. A standard SINR model can be formulated as follows, which states that a message sent by u is correctly received by v if and only if

$$SINR(u, v) = \frac{P_u \cdot d(u, v)^{-\alpha}}{\mathcal{N} + \sum_{w \in S \setminus \{u\}} P_w \cdot d(w, v)^{-\alpha}} \geq \beta \tag{1.1}$$

holds, where $\alpha \in (2, 6]$ is the path-loss exponent, threshold $\beta > 1$ is determined by hardware, $P_u \cdot d(u, v)^{-\alpha}$ is the received signal power of node v from u, $\mathcal{N}$ is the ambient noise, $P_w \cdot d(w, v)^{-\alpha}$ is the received signal power of node v from w, and $S \subseteq V$ denotes the set of nodes transmitting simultaneously with u.

Adversary Model In consensus algorithms, nodes can be classified into two categories: honesty nodes and faulty nodes. Honesty nodes strictly follow the protocol procedures, while faulty nodes may exhibit one of the six types of node failures mentioned in Sect. 1.1.1.2 (Failure Model). Although there are six potential failure types, consensus algorithms in practical wired networks typically consider only two categories: crash failures and Byzantine failures. Regardless of the error type, wired network consensus algorithms generally do not account for adversary attacks on the network layer, especially when considering Byzantine failures. This is due to the stable transmission and advanced protocols in wired networks, which make adversary attacks on the network layer easily identifiable and defendable.

However, consensus algorithms in wireless networks cannot ignore adversary attacks on the network layer. Due to the open nature of wireless network channels, jamming attacks are a common form of attack in wireless networks. Although jamming attacks can be mitigated through wireless network protocols, many consensus layer algorithms are designed to incorporate mechanisms to resist jamming attacks, thereby improving consensus efficiency. Additionally, omission faults refer to packet loss errors caused by network dynamics. In wired consensus, such errors are generally not considered separately and are usually encompassed by synchrony abstraction. Communication protocols in wired networks also handle these errors through retransmission. However, in wireless consensus, this must be carefully addressed, as increased retransmission leads directly to higher energy consumption. Many devices in wireless scenarios are battery-powered, need to operate for extended periods, and cannot support algorithms with high energy consumption.

1.2.3.2 Impossibility Result

In distributed systems, whether in wired or wireless networks, achieving consensus faces various impossibility issues, making the design of effective consensus algorithms highly challenging.

- **CAP Theorem:** Both wired and wireless networks confront the difficulties outlined by the CAP theorem, which reveals that it is impossible for any distributed system to simultaneously ensure consistency, availability, and partition tolerance. Network partitions may force a choice between consistency and availability. While partitions are less common in wired networks, they must still be considered for system response under extreme conditions. Wireless networks, inherently unstable, encounter network partitions more frequently, making it harder to maintain consistency and availability.
- **FLP Impossibility Result:** The FLP [44] impossibility result states that in asynchronous networks if even a single node fails, it is impossible to guarantee that all non-failing nodes will reach consensus. This result is particularly pertinent in wired networks where infrastructure is reliable, yet node failures or network delays can still hinder consensus. In wireless networks, the result is even more pronounced due to the instability and higher failure rates of nodes. The asynchronous nature and instability of wireless networks further complicate consensus achievement.

In achieving consensus, wireless networks face not only asynchronous system issues similar to those in wired networks but also unique limitations and challenges inherent to wireless communication.

- **Unreliable Signals:** Wireless network signals are susceptible to physical obstructions like buildings and weather, leading to frequent communication failures. Additionally, the common use of broadcast methods for data transmission in wireless networks increases the probability of data collisions and errors, further affecting message reliability and overall system performance.
- **Energy Limits:** Wireless devices such as sensors and mobile devices are typically limited by strict energy constraints, restricting their operational capabilities and continuous communication duration. The need to conserve energy means these devices cannot participate consistently in complex computations and data exchanges, posing significant challenges to the implementation of consensus algorithms.
- **Dynamic Network:** Wireless devices may move, causing frequent changes in network topology. This high level of dynamicity in the network environment complicates maintaining a stable network view and synchronizing state among nodes, increasing the complexity of correctly executing consensus algorithms.
- **Security Issues:** Wireless networks are more susceptible to security attacks, such as man-in-the-middle attacks and signal jamming. These security threats can lead to data breaches and directly impact the integrity and accuracy of the consensus

process. Verifying node identity and encrypting data are critical to ensuring the reliability of consensus in wireless networks.

1.2.3.3 Research Directions

Wired and wireless networks have different focuses and strategies in the design of consensus algorithms. Wired network consensus algorithms emphasize high performance, low latency, robust fault tolerance, and a certain level of energy efficiency. Here are several key aspects of wired network consensus algorithm design:

- **Performance:** Wired network consensus algorithms highly focus on message complexity, latency, and throughput. Due to the higher bandwidth and stability of wired networks, they can handle high message complexity while maintaining low latency and high throughput. For example, low latency and high throughput are crucial in high-frequency trading and financial systems.
- **Scalability:** Consensus algorithms in wired networks need to maintain performance and efficiency as the number of nodes increases. By using sharding techniques, the network can be divided into multiple subnetworks, each independently handling the consensus process, thus enhancing the system's scalability. This approach ensures that as the network grows, the system's performance does not significantly degrade.
- **Fault Tolerance:** Wired network consensus algorithms use BFT consensus algorithms to ensure the system continues to function correctly in the presence of faulty or malicious nodes. These algorithms can tolerate a portion of nodes failing or behaving maliciously, maintaining the system's consistency and reliability. For example, PBFT is a commonly used BFT algorithm widely applied in distributed systems.
- **Energy Efficiency:** Although energy efficiency is not the primary concern in wired networks, with the increasing demand for green computing and sustainability, consensus algorithm design is gradually shifting from high-energy consumption Proof-of-Work (PoW) to more energy-efficient Proof-of-Stake (PoS). The PoS mechanism achieves consensus by staking tokens, reducing the consumption of computational resources and thus improving energy efficiency.

Wireless network consensus algorithms, on the other hand, focus more on low energy consumption, strong interference resistance, and adaptability to dynamic environments. Here are several key aspects of wireless network consensus algorithm design:

- **Energy Efficiency:** Devices in wireless networks typically rely on battery power, so consensus algorithms must minimize energy consumption. Using low-power communication techniques such as duty cycling and preamble sampling, along with low-power hardware design, can significantly extend the battery life

of devices. Optimizing the algorithm to complete the consensus process with minimal energy consumption is a core goal in wireless network design.
- **Interference Resistance:** Wireless network environments are susceptible to external interference, so consensus algorithms must have strong interference resistance. Techniques such as spectrum sensing, dynamic frequency selection, and redundancy design can enhance system reliability. For example, spectrum sensing allows devices to detect and avoid interference bands, while dynamic frequency selection enables devices to communicate on less-interfered frequency bands.
- **Scalability:** Although wireless networks have limited bandwidth and resources, sharding techniques can divide the network into multiple subnetworks, each independently handling consensus tasks, thereby increasing processing capacity and conserving energy. This approach not only improves system scalability but also enhances overall network performance.
- **Adaptive Algorithms:** The dynamic nature of wireless networks requires consensus algorithms to adapt to network conditions. By dynamically adjusting protocol parameters, algorithms can maintain efficient operation under varying network conditions, ensuring the smooth progression of the consensus process.

1.3 Challenges Faced by Wireless Consensus

In wireless consensus, challenges arise from device characteristics, wireless channel features, and diversity of wireless networks.

1.3.1 Resource-Constrained Device

Wireless networks are often deployed in dynamic environments, where devices are typically powered by batteries. This is particularly true in wireless ad hoc and some low-power wireless networks. Resource limitations manifest in three areas: communication, computing, and storage.

- **Communication:** The spectrum and bandwidth in wireless networks are limited. The scarcity of spectrum resources has always been a significant challenge in wireless communications. Hence, wireless consensus algorithms cannot establish separate channels for any two points as in wired networks. Typically, multiple or even all nodes share the same channel. Innovative spectrum allocation strategies, like spectrum sharing, spectrum sensing, and dynamic spectrum allocation, are needed to manage the limited spectrum resources more effectively and improve wireless consensus efficiency.
- **Computation:** IoT devices often have limited processing capabilities, which is particularly evident in some devices. For example, processor speeds often do

not exceed 1 GHz, as observed in ARM-M series chips. Limited computational power makes it difficult to deploy certain cryptographic tools in IoT devices, such as threshold signatures and encryption used in asynchronous wired network consensus algorithms like HBBFT. These heavier cryptographic tools consume a lot of energy, making them unsuitable for low-power wireless networks.

- **Storage:** The storage capacity of IoT devices is also very limited. For instance, some development boards might only have a few MBs of total storage resources, like the Nucleo144-STM32F767 equipped with only 2 MB of Flash storage and 512 KB of SRAM. This limited storage space is a challenge for wireless consensus that requires storing large amounts of data. Therefore, reliance on external storage solutions or data compression and optimization strategies might be necessary to save space. Additionally, developers need to choose efficient data structures and storage mechanisms to reduce memory usage and improve storage efficiency.

1.3.2 Overcoming the Effects of Wireless Channels

The instability of wireless channels presents multiple challenges to consensus algorithms. These challenges include, but are not limited to, the following aspects:

- **Open Channels:** Wireless communication uses public spectrums, with multiple devices and networks operating in the same frequency band. This leads to channel congestion and interference issues, as different devices compete for the same frequency. This competition increases latency and decreases throughput in wireless consensus. Intelligent spectrum management and collaborative mechanisms are needed to ensure fair and efficient sharing of spectrum resources, reduce interference, and improve network performance.
- **Competitive Interference and Blocking:** In situations where spectrum resources are limited, multiple devices competing for the same channel can lead to data collisions, reducing the success rate of data transmissions. This interference and blocking issue is especially significant in high-density wireless networks. Intelligent collision avoidance and backoff algorithms, as well as appropriate quality of service (QoS) strategies, are needed to optimize the use of network resources, improve wireless consensus throughput, and reduce latency.
- **Fading Channels:** Wireless signals are prone to instability due to obstacles, reflections, and multipath propagation during transmission. This can lead to packet loss and retransmission, reducing the reliability and efficiency of data transmission. To address the challenges of fading channels, adaptive modulation and coding techniques, smart antenna systems, and advanced channel estimation algorithms are needed to enhance signal stability and transmission quality, thereby increasing wireless consensus efficiency.
- **Dynamism:** The mobility in wireless environments increases the instability of connections. The high mobility of nodes like drones and mobile vehicles

causes frequent changes in network topology, impacting network stability and reliability. Challenges include maintaining connection stability, ensuring path reliability, and effectively conducting neighbor discovery and management, and energy management in high mobility conditions. To address these issues, communication protocols capable of quickly adapting to network changes, real-time routing information updates, robust data transmission mechanisms, and efficient energy management strategies are required.

Compared to wired networks, the most distinctive feature of wireless networks is their diverse environments. Different wireless networks require specifically designed consensus algorithms. If a consensus algorithm from one wireless environment is directly transferred to another, it can lead to efficiency and security issues. Different communication protocols and underlying network technologies exhibit varying characteristics in terms of spectrum utilization, signal propagation, and interference management. The frequency usage and transmission power of different protocols directly affect the level of interference between devices. For instance, some short-range communication protocols might operate in the same frequency band, leading to mutual interference, especially in high-density device deployments. In such cases, it is essential to establish anti-interference and spectrum coordination mechanisms to minimize interference and ensure communication reliability, thereby enhancing consensus efficiency. The dynamic nature of network topology is also a key issue. The mobility of devices and their joining/leaving the network cause constant changes in the network structure. Different communication protocols have varying degrees of adaptability. For example, in some wireless networks, frequent node mobility may require consensus algorithms to quickly adapt to the changing network topology to ensure consistency and efficiency. In other networks, nodes may be relatively static but still need to address potential communication disruptions and interference problems.

Therefore, the design of wireless consensus algorithms must fully consider these characteristics and adopt adaptive mechanisms to address different challenges. Spectrum sensing and dynamic frequency selection techniques can effectively reduce channel interference and improve communication quality. Spectrum coordination technologies can manage the spectrum usage of multiple devices to avoid mutual interference and enhance overall network efficiency. To adapt to the constantly changing network topology, consensus algorithms need to have the ability to quickly adjust, ensuring the system can operate efficiently in dynamic environments.

1.4 Summary

This chapter provides a brief overview of wireless consensus. We introduce the fundamentals of wireless consensus. Wireless networks and consensus algorithms are the basic components of wireless consensus. Similar to the development of most algorithms, the history of wireless consensus has gone through a journey from theory to practice. In the later stages of wireless consensus development, researchers have increasingly focused on its performance in real-world systems. This chapter also summarizes the simulators and practical testbeds currently used by researchers. Additionally, we compare wireless consensus protocols with wired consensus protocols. Wired consensus protocols have developed earlier and are more mature than wireless consensus protocols. Wireless networks exhibit unique characteristics compared to wired networks, prompting us to compare them across three layers: physical layer, network layer, and consensus layer. Finally, we summarize some of the challenges faced by wireless consensus, primarily resource-constrained devices and the instability of wireless channels.

References

1. Lee, J.-S., Su, Y.-W., & Shen, C.-C. (2007). A comparative study of wireless protocols: Bluetooth, UWB, ZigBee, and Wi-Fi. In *IECON 2007-33rd Annual Conference of the IEEE Industrial Electronics Society* (pp. 46–51). IEEE.
2. Mekki, K., Bajic, E., Chaxel, F., & Meyer, F. (2019). A comparative study of LPWAN technologies for large-scale IoT deployment. *ICT Express, 5*(1), 1–7.
3. Dwork, C., Lynch, N., & Stockmeyer, L. (1988). Consensus in the presence of partial synchrony. *Journal of the ACM (JACM), 35*(2), 288–323.
4. Abraham, I., Malkhi, D., Nayak, K., Ren, L., & Yin, M. (2020). Sync hotstuff: Simple and practical synchronous state machine replication. In *2020 IEEE Symposium on Security and Privacy (SP)* (pp. 106–118). IEEE.
5. Castro, M., Liskov, B., et al. (1999). Practical byzantine fault tolerance. In *OsDI* (Vol. 99, pp. 173–186).
6. Miller, A., Xia, Y., Croman, K., Shi, E., & Song, D. (2016). The honey badger of bft protocols. In *Proceedings of the 2016 ACM SIGSAC Conference on Computer and Communications Security* (pp. 31–42).
7. Dziembowski, S., Faust, S., Kolmogorov, V., & Pietrzak, K. (2015). Proofs of space. In Gennaro, R. & Robshaw, M. (Eds.), *Advances in Cryptology – CRYPTO 2015* (pp. 585–605). Springer.
8. Chen, L., Xu, L., Shah, N., Gao, Z., Lu, Y., & Shi, W. (2017). On security analysis of proof-of-elapsed-time (poet). In *International Symposium on Stabilization, Safety, and Security of Distributed Systems* (pp. 282–297). Springer.
9. Bowers, K. D., Juels, A., & Oprea, A. (2009). Proofs of retrievability: Theory and implementation. In *Proceedings of the 2009 ACM Workshop on Cloud computing Security* (pp. 43–54)
10. Miller, A., Juels, A., Shi, E., Parno, B., & Katz, J. (2014). Permacoin: Repurposing bitcoin work for data preservation. In *2014 IEEE Symposium on Security and Privacy* (pp. 475–490). IEEE.
11. What is delegated proof of stake? https://crypto.com/university/what-is-dpos-delegated-proof-of-stake
12. De Angelis, S., Aniello, L., Baldoni, R., Lombardi, F., Margheri, A., & Sassone, V. (2018). PBFT vs proof-of-authority: Applying the CAP theorem to permissioned blockchain. In *Italian Conference on Cyber Security (06/02/18)*.
13. Seth, S. Proof of activity. https://www.investopedia.com/terms/p/proof-activity-cryptocurrency.asp

14. Gai, F., Wang, B., Deng, W., & Peng, W. (2018). Proof of reputation: A reputation-based consensus protocol for peer-to-peer network. In *International Conference on Database Systems for Advanced Applications* (pp. 666–681). Springer.
15. Karantias, K., Kiayias, A., & Zindros, D. (2020) Proof-of-burn. In *International Conference on Financial Cryptography and Data Security* (pp. 523–540). Springer.
16. Xu, M., Zhao, F., Zou, Y., Liu, C., Cheng, X., & Dressler, F. (2022). BLOWN: A blockchain protocol for single-hop wireless networks under adversarial SINR. *IEEE Transactions on Mobile Computing, 22*(8), 4530–4547.
17. Xu, M., Liu, C., Zou, Y., Zhao, F., Yu, J., & Cheng, X. (2021). wChain: A fast fault-tolerant blockchain protocol for multihop wireless networks. *IEEE Transactions on Wireless Communications, 20*(10), 6915–6926.
18. Zou, Y., Xu, M., Yu, J., Zhao, F., & Cheng, X. (2021). A fast consensus for permissioned wireless blockchains. *IEEE Internet of Things Journal, 10*(14), 12102–12111.
19. Slimcoin. http://slimco.in/
20. Kuhn, F., Lynch, N., & Newport, C. (2011). The abstract mac layer. In *International Symposium on Distributed Computing* (pp. 48–62). Springer.
21. Newport, C. (2014). Consensus with an abstract mac layer. In *Proceedings of the 2014 ACM Symposium on Principles of Distributed Computing* (pp. 66–75).
22. Newport, C, & Robinson, P. (2018). Fault-tolerant consensus with an abstract mac layer. arXiv preprint arXiv:1810.02848.
23. Yu, D., Zhang, Y., Huang, Y., Jin, H., Yu, J., & Hua, Q.-S. (2018). Exact implementation of abstract mac layer via carrier sensing. In *IEEE INFOCOM 2018-IEEE Conference on Computer Communications* (pp. 1196–1204). IEEE.
24. Chockler, G., Demirbas, M., Gilbert, S., Newport, C., & Nolte, T. (2005). Consensus and collision detectors in wireless ad hoc networks. In *Proceedings of the Twenty-Fourth Annual ACM Symposium on Principles of Distributed Computing* (pp. 197–206).
25. Scutari, G., & Barbarossa, S. (2008). Distributed consensus over wireless sensor networks affected by multipath fading. *IEEE Transactions on Signal Processing, 56*(8), 4100–4106.
26. Aysal, T. C., Sarwate, A. D., & Dimakis, A. G. (2009). Reaching consensus in wireless networks with probabilistic broadcast. In *2009 47th Annual Allerton Conference on Communication, Control, and Computing (Allerton)* (pp. 732–739). IEEE.
27. Borran, F., Prakash, R., & Schiper, A. (2008). Extending paxos/lastvoting with an adequate communication layer for wireless ad hoc networks. In *2008 Symposium on Reliable Distributed Systems* (pp. 227–236). IEEE.
28. Vasudevan, S., DeCleene, B., Immerman, N., Kurose, J., & Towsley, D. (2003). Leader election algorithms for wireless ad hoc networks. In *Proceedings DARPA Information Survivability Conference and Exposition* (Vol. 1, pp. 261–272). IEEE.
29. Dong, Q., & Liu, D. (2009). Resilient cluster leader election for wireless sensor networks. In *2009 6th Annual IEEE Communications Society Conference on Sensor, Mesh and Ad Hoc Communications and Networks* (pp. 1–9). IEEE.
30. Raychoudhury, V., Cao, J., & Wu, W. (2008). Top k-leader election in wireless ad hoc networks. In *2008 Proceedings of 17th International Conference on Computer Communications and Networks* (pp. 1–6). IEEE.
31. Gołębiewski, Z., Klonowski, M., Koza, M., & Kutyłowski, M. (2009). Towards fair leader election in wireless networks. In *International Conference on Ad-Hoc Networks and Wireless* (pp. 166–179). Springer.
32. Richa, A., Scheideler, C., Schmid, S., & Zhang, J. (2011). Self-stabilizing leader election for single-hop wireless networks despite jamming. In *Proceedings of the Twelfth ACM International Symposium on Mobile Ad Hoc Networking and Computing* (pp. 1–10).
33. Al Nahas, B., Duquennoy, S., & Landsiedel, O. (2017). Network-wide consensus utilizing the capture effect in low-power wireless networks. In *Proceedings of the 15th ACM Conference on Embedded Network Sensor Systems* (pp. 1–14).
34. Poirot, V., Al Nahas, B., & Landsiedel, O. (2019). Paxos made wireless: Consensus in the air. In *EWSN* (pp. 1–12).

35. Goyal, H., Saha, S., et al. (2022). Reli: Real-time lightweight byzantine consensus in low-power iot-systems. In *2022 18th International Conference on Network and Service Management (CNSM)* (pp. 275–281). IEEE.
36. Xu, W., & Kapitza, R. (2018). Ratcheta: Memory-bounded hybrid byzantine consensus for cooperative embedded systems. In *2018 IEEE 37th Symposium on Reliable Distributed Systems (SRDS)* (pp. 103–112). IEEE.
37. Xu, W., Wegner, M., Wolf, L., & Kapitza, R. (2017). Byzantine agreement service for cooperative wireless embedded systems. In *2017 47th Annual IEEE/IFIP International Conference on Dependable Systems and Networks Workshops (DSN-W)* (pp. 10–15). IEEE.
38. Moniz, H., Neves, N. F., & Correia, M. (2010). Turquois: Byzantine consensus in wireless ad hoc networks. In *2010 IEEE/IFIP International Conference on Dependable Systems & Networks (DSN)* (pp. 537–546). IEEE.
39. Zhou, Z., Onireti, O., Zhang, L., & Imran, M. A. (2021). Performance analysis of wireless practical byzantine fault tolerance networks using IEEE 802.11. In *2021 IEEE Globecom Workshops (GC Wkshps)* (pp. 1–6). IEEE.
40. Xu, M., Zhao, F., Zou, Y., Liu, C., Cheng, X., & Dressler, F. (2022). BLOWN: A blockchain protocol for single-hop wireless networks under adversarial SINR. *IEEE Transactions on Mobile Computing, 22*(8), 4530–4547.
41. Xu, M., Liu, C., Zou, Y., Zhao, F., Yu, J., & Cheng, X. (2021). wChain: A fast fault-tolerant blockchain protocol for multihop wireless networks. *IEEE Transactions on Wireless Communications, 20*(10), 6915–6926.
42. Bhat, A., Bandarupalli, A., Nagaraj, M., Bagchi, S., Kate, A., & Reiter, M. K. (2023). Eesmr: Energy efficient bft—smr for the masses. In *Proceedings of the 24th International Middleware Conference* (pp. 1–14).
43. Böhm, H., Distler, T., & Wägemann, P. (2024). Tinybft: Byzantine fault-tolerant replication for highly resource-constrained embedded systems. In *Proceedings of the 30th IEEE Real-Time and Embedded Technology and Applications Symposium (RTAS'24).*
44. Fischer, M. J., Lynch, N. A., & Paterson, M. S. (1985). Impossibility of distributed consensus with one faulty process. *Journal of the ACM (JACM), 32*(2), 374–382.

Chapter 2
Fault-Tolerant Wireless Consensus

Fault-tolerant wireless consensus systems face numerous challenges in practical applications, among which unstable communication poses a pivotal issue. Some prior works achieve consensus among all participating nodes by relying on a dependable wireless channel, e.g., [2] and [3]. In contrast, real-world wireless networks contend with an unreliable and error-prone shared wireless channel. In a real wireless setting, the reliability and performance of the system can easily be affected by factors such as signal interference, node movement, device failures, insufficient signal coverage, and network congestion. These sources of unstable communication could lead to data transmission delays, increased packet loss, and even communication breakdowns, severely impacting the data consistency, real-time nature, and overall coordination capability of the system.

The adverse effects of unstable communication on fault-tolerant wireless consensus systems are multifaceted. First, communication instability can cause inconsistencies in data among nodes, disrupting a system's overall decision-making and coordination capacity. For instance, in a drone swarm, data inconsistency might lead drones to execute divergent tasks, increasing the risk of task failure. In a smart grid, data inconsistency could result in unbalanced loads, heightening the risk of power outages. Second, unstable communication can increase system latency, affecting its real-time capabilities. In scenarios requiring rapid response, such as disaster rescue and smart traffic, latency increases might prevent the system from responding in time, leading to serious consequences. Furthermore, unstable communication can cause resource waste, particularly in devices with limited energy such as sensors and drones, where repeated data transmission can increase communication costs, deplete device power more quickly, and affect task sustainability. Ultimately, unstable communication can weaken the system's robustness, making it more vulnerable to single-point failures. For example, in smart agriculture, the offline state of certain sensors may lead to incomplete environmental data, affecting the accuracy of decisions; in smart grids, the failure of certain nodes can cause a decrease in the stability of the whole grid.

M. Xu et al., *Wireless Consensus*, Wireless Networks,
https://doi.org/10.1007/978-3-031-70859-6_2

Considering the sizable influence of unstable communication on wireless fault-tolerant consensus systems, it is vital to recognize and tackle this problem to bolster the system's reliability and efficiency. This segment will examine unstable wireless communication in depth, highlighting its effects on the performance of fault-tolerant consensus systems. We have presented the paramount communication method in fault-tolerant consensus-atomic multicast. We subsequently discern that the primary hindrance for traditional fault-tolerant consensus algorithms, preventing them from operating effectively in unstable wireless networks, stems from their incapacity to execute reliable multicast in such conditions. We then suggest a plausible effective protocol that names the abstract Media Access Control (MAC) layer, offering an overview of its feasibility along with particular instances.

2.1 Impacts of Unstable Communication

Wireless fault-tolerant consensus systems play a crucial role in ensuring reliable decision-making and agreement among nodes in wireless networks. However, these systems face various challenges due to the inherent characteristics of wireless communication. Unstable communication can significantly impact the performance and effectiveness of wireless fault-tolerant consensus protocols. In this section, we will explore the different types of wireless networks used for fault-tolerant consensus, focusing on Wi-Fi and Bluetooth technologies. We will also discuss the causes of unstable communication, including wireless signal interference, node activity, device failures, and network congestion. By understanding these factors and their impact on wireless fault-tolerant consensus systems, we can develop strategies to mitigate their effects and improve the overall reliability and robustness of these systems in mobile wireless networks.

2.1.1 Characteristics of Fault-Tolerant Wireless Networks

As shown in Table 2.1, wireless local area networks (WLANs) are constructed as an alternative to wired local area networks (LANs). They aim to facilitate connectivity for mobile devices or simply to eliminate the necessity for a wired infrastructure to connect computers within homes and office buildings to each other and the Internet. They are now extensively adopted in various formats of the IEEE 802.11 standard (Wi-Fi), delivering bandwidths of 10–100 Mbps, covering distances up to 1.5 kilometers.

Wireless metropolitan area networks (WMANs) follow the IEEE 802.16 WiMAX standard. This type of network is aimed to offer an alternative to wired connections for homes and office buildings and to surpass 802.11 Wi-Fi networks in certain applications. Wireless wide area networks (WWANs) form the basis of most mobile phone networks, as they utilize digital wireless network technologies

Table 2.1 Comparison of different network types

Type	Example	Range	Bandwidth (Mbps)	Latency (ms)
LAN	Ethernet	1–2 km	10–10,000	1–10
WAN	IP routing	Worldwide	0.010–600	100–500
MAN	ATM	2–50 km	1–600	10
Internet	Internet	Worldwide	0.5–600	100–500
WPAN	IEEE 802.15.1	10–30 m	0.5–2	5–20
WLAN	Wi-Fi (IEEE 802.11)	0.15–1.5 km	11–108	5–20
WMAN	WiMAX (IEEE 802.16)	5–50 km	1.5–20	5–20
WWAN	5G phone	Cell: 1–5 km	348–14.4	100–500

such as the Global System for Mobile Communication (GSM) standard, which is popularly used across numerous countries globally. Designed to operate across extensive spheres, typically across entire nations or continents, these networks employ cellular radio connections and thus provide wide area mobile Internet connectivity for portable devices. Despite the typically low data rates offered by the mentioned cellular networks from 9.6 to 33 kbps, the "third generation" (3G) of mobile phone networks is now available offering data transmission rates in the range of 2–14.4 Mbps while stationary and 348 kbps while in motion (such as in a car). The underlying technology is identified as the Universal Mobile Telecommunications System (UMTS). A roadmap has been clearly established to evolve UMTS toward 4G data rates of up to 100 Mbps.

2.1.2 Wireless Signal Interference

Wireless signal interference arises from a multitude of sources, which include environmental noise, other wireless devices, multipath interference, and spectrum congestion, resulting in unpredictable message loss. Each of these problems can severely impact the performance and reliability of wireless communication systems, particularly those relying on fault-tolerant consensus mechanisms. This can pose challenges in achieving consensus since nodes may not receive all essential messages [4].

Environmental noise encompasses both natural and artificial electromagnetic noise. Natural sources such as thunder, lightning, and solar radiation, as well as artificial sources like industrial machinery and household appliances, contribute to this type of interference. These elements can lead to significant signal attenuation and data packet loss, thus impeding effective communication.

Interference from other wireless devices is another critical factor. Devices such as Wi-Fi routers, Bluetooth peripherals, and microwave ovens often operate within the same frequency bands, leading to signal conflicts. These conflicts manifest as data transmission errors or delays, which can severely compromise the integrity of communication systems.

Multipath interference occurs when wireless signals encounter reflections, refractions, and scatterings during transmission. These interactions cause variations in signal phase and amplitude, which result in interference that degrades signal clarity and the accuracy of data transmission. The phenomenon of multipath interference is particularly problematic in urban environments with numerous reflective surfaces.

Spectrum congestion refers to the overcrowding of available frequency bands, especially in areas with a high density of wireless devices. This congestion leads to channel conflicts and increased data transmission delays as multiple devices vie for the same spectrum resources.

The detrimental effects of signal interference on fault-tolerant wireless consensus systems are multifaceted, impacting four primary areas: data transmission accuracy, communication latency, and system robustness.

- **Data Transmission Errors.** Signal interference can corrupt or cause the loss of data packets, thereby preventing receiving nodes from correctly decoding the information. This undermines the accuracy and reliability of consensus algorithms. In practical applications such as drone swarms and smart grids, this could result in drones failing to execute tasks correctly or cause erroneous load distributions within the grid.
- **Increased Communication Delays.** Interference necessitates multiple retransmissions of data packets, which prolongs communication times and impacts real-time system performance. In critical sectors like disaster relief and smart farming, such delays could hinder timely rescue operations or affect crop management and growth.
- **Decreased System Robustness.** Interference weakens overall communication quality, making the system more vulnerable to single-point failures. In applications like smart grids and smart agriculture, this can lead to reduced grid stability and incomplete environmental data collection, respectively.
- **Energy Efficiency.** Signal interference often forces devices to retransmit data or increase transmission power, leading to higher energy consumption. In wireless networks, where many devices rely on battery power or have limited energy resources, this can drastically reduce the network's lifespan and the ability of the devices to participate in consensus. Interference can cause repeated communication attempts, increasing overhead and wasting precious energy resources. As devices consume more energy, they may fail or become unavailable for consensus, further compromising the system's overall reliability and fault tolerance. Designing energy-efficient protocols that can mitigate interference is thus essential for ensuring the long-term sustainability of fault-tolerant wireless networks.

The impact of wireless signal interference on fault-tolerant wireless consensus systems is profound and multifaceted. It not only increases the incidence of data transmission errors and communication delays but also escalates energy consumption and diminishes system robustness. Addressing these challenges necessitates the implementation of advanced interference management techniques and the enhancement of system design to bolster the reliability and efficiency of wireless communications.

2.1.3 The Detrimental Effects of Unstable Communication

2.1.3.1 Impacts of Node Activity

The dynamic nature of nodes in mobile wireless networks is a pivotal factor that incites communication inconsistency. This dynamicity encompasses various behaviors, such as the entry and exit of nodes, dormancy, and awakening. In numerous wireless networks, nodes exhibit mobility. This mobility can result in swiftly evolving network topologies, posing challenges to sustaining uniform states among nodes [5]. The following part offers an in-depth examination of these behaviors and their subsequent influence on the performance and reliability of wireless fault-tolerant consensus systems.

Node admission refers to the inclusion of a new node into the network, which begins communication and data exchange with the existing nodes. The inclusion of a fresh node modifies the current network topology, impacting the data transmission channels and load on the network. Furthermore, the new node requires time for initialization, which incorporates the discovery of adjacent nodes, data synchronization, and consensus participation. These activities can potentially result in short-term communication latencies. The new node must also ascertain the current system status and data to ensure data consistency with the other nodes; if not managed appropriately, this can lead to data inconsistency. Additionally, the introduction of a new node can potentially exacerbate competition for limited resources such as bandwidth and spectrum, elevating the complexity of resource management.

Node departure refers to the removal of a node from the network due to factors like failure, power depletion, or voluntary withdrawal. The removal of a node can result in delayed data transmission to other nodes, which can lead to data loss. The remaining nodes must remodify the network topology and data transmission paths, thus potentially causing short-term communication disruptions and latencies. If a departing node plays a crucial role in the consensus process, its removal can either disrupt the consensus process or necessitate its restart. Frequent node exits can destabilize and reduce the overall dependability of the system, thereby augmenting the complexity of network maintenance.

Node dormancy refers to a node transitioning into a low-energy state, thereby temporarily suspending communication and data processing to conserve energy. Dormant nodes are incapable of transmitting or receiving data, resulting in communication link interruptions. When awakened, the dormant nodes need to update their data, which could potentially lead to data lag or inconsistency. Recurrent entries into and exits from the dormant state can affect the system's reliability, especially in scenarios involving critical tasks. Despite dormancy being an effective energy-saving mechanism, repeated dormancy and awakenings escalate the complexity of energy management.

Node awakening refers to a node reverting to its normal operation state from a dormant state, thereby reinitiating network communication and data processing. Once awakened, nodes require time to synchronize data and reestablish communication links, causing short-term syncing delays. Moreover, awakening nodes might cause abrupt network load fluctuations, thus heightening the instant load on the network. Awakened nodes need to re-sync data with other nodes to ensure data consistency; untimely syncing might result in data conflicts and decision-making errors. Frequent awakenings and dormant states might inflict dynamic changes in the network topology, affecting network connectivity and coverage range.

The dynamic behavior of nodes can engender issues such as data inconsistency, communication delay, or disruption, critically impacting the reliability and performance of wireless fault-tolerant consensus systems. Data inconsistency could threaten the precision and dependability of consensus algorithms, potentially leading to decision-making errors and system failures in applications, including smart agriculture and smart grids. Communication delays and disruptions might compromise the real-time performance and stability of the system, which can subsequently lead to mission failure or rescue delay in scenarios such as drone clusters.

2.1.3.2 Impacts of Device Failure

Within the realm of wireless fault-tolerant consensus systems, equipment failure plays a substantial role that significantly impacts communication stability as well as the overall reliability of the system. A primary focus could be issued surrounding failures that originate from the equipment itself. These encompass hardware failures, issues with the software, as well as power depletion. More specifically, hardware failures encapsulate occurrences when the physical components of the device, such as circuitry or peripherals, fail as a result of damage, decay resulting from age, or even defects in the manufacturing process. The subsequent interference in communication, potential loss of vital data, along a marked reduction in system stability, all serve as glaring examples of consequences following a hardware failure. It is worth noting that this section primarily focuses on equipment faults that are instigated by the equipment itself; these include hardware and software faults along with power exhaustion. Faults induced due to Byzantine activity such as malicious attacks or misbehavior by nodes will be discussed in the sections to follow. It is also important to mention that although node departure does share certain commonalities with equipment failure, it can be caused by reasons not related to faults, such as voluntary exit. Hence, it is noteworthy to distinguish and analyze these causes separately.

Software failure is one of the key challenges in device failure. The factors that lead to software failure are complex and multifarious, potentially stemming from bugs at the code level, such as buffer overflows, dangling pointers, and race conditions. They could also be the result of software architecture design defects, such as excessively high module coupling, unclear interface definitions, or imperfect exception-handling mechanisms. In addition, changes in the operating environment, for example, memory fragmentation, intensifying resource competition, and updates or iterations in third-party libraries, can also cause software failures. Such failures can not only interrupt or anomalize node services but also trigger domino effects, such as cascading failures or inconsistent data, threatening the system's availability and integrity.

Take the example of a drone swarm collaborative task. Suppose the implementation of the formation control algorithm exhibits flaws such as improper handling of boundary conditions or defects in concurrency control. In that case, it may cause drones to suddenly go out of control, deviate from their preset trajectories, and trigger collision accidents, resulting in task failures and equipment damage. This not only jeopardizes the overall performance of the drone swarm but also brings additional economic losses and safety risks. Another example is in tasks for gathering data and estimating state parameters in a smart grid. If key algorithm implementations, such as signal processing, feature extraction, and data fusion, on nodes are defective, it may result in errors in power parameter estimations, distortions in grid topology identification, and inaccurate fault diagnosis results, affecting the scientific nature and reliability of grid scheduling.

Device failure due to power depletion is as important as software failure. In actual deployments of wireless fault-tolerant networks, nodes are often located in the wilderness, where it is difficult to connect to stable power supplies, and they must rely on a limited battery supply. As these nodes continuously perform communication, computing, and storage tasks, their energy is constantly depleted. If energy resupply is delayed or load distribution is uneven, some nodes may lose their connection too soon due to energy exhaustion. Wireless nodes are frequently battery powered and must conserve energy. Since consensus algorithms can be communication intensive, the design of energy-efficient non-Byzantine fault-tolerant consensus (NBFTC) algorithms poses a significant challenge [6]. The drop in node density leaves blind spots in the network coverage, changing the network topology, disabling routing mechanisms, and reducing the system's fault tolerance capacity. Simultaneously, lost connections can prevent the timely uploading of locally stored critical data, affecting data integrity and consistency.

In the scenario of smart agriculture applications, sensor networks are typically deployed in remote, harsh conditions in the fields and are difficult to maintain frequently. If large numbers of nodes in a network lose their connection due to energy consumption, extensive data collection blind spots can occur. Critical parameters, such as crop development parameters and soil environment parameters, become impossible to monitor in real time, impacting the timeliness and accuracy of intelligent planting decisions and causing incalculable losses to crop growth and agricultural production.

The dynamic behavior of the nodes is inherently connected to the potential for equipment failure and, as such, requires a holistic approach in its consideration. As nodes are introduced into the network, procedural checks need to be carried out upon both the hardware and software to assuage any doubts concerning possible faults and ensure functionality. Furthermore, it is also important to check the battery levels of newly introduced nodes to circumvent the occurrence of premature failures. Instances of node departures could potentially be triggered by inherent equipment failures such as noted hardware failures or instances of power depletion. Moreover, it becomes imperative to quickly detect and record the causal factors behind a node's departure. This is to facilitate both the analysis of failures and streamline maintenance procedures.

When a node undergoes a sleep and wake cycle, the state of the hardware as well as the software needs to be stable to prevent wake failures due to underlying faults. A vital consideration is also to ensure that there is adequate battery power to support normal operations post-awakening, thereby preventing failures that could stem from power depletion.

The fallout from equipment failures manifests itself in serious issues such as the disruption of communication, loss of data, and incrementing system maintenance costs. These issues significantly affect the dependability and performance metrics of wireless fault-tolerant consensus systems. Disruptions in communication can have a potentially detrimental effect on the system's real-time information processing and overall stability, potentially causing task failure or delaying the time-sensitive collection and transmission of data, which is crucial for applications in smart agriculture and drone swarm operations. Data loss is yet another issue to be considered as it can greatly interfere with the algorithm's accuracy and reliability, potentially resulting in massive decision-making errors or incomplete data uploads. This is especially crucial in the context of smart grids and intelligent transportation systems.

2.1.3.3 Impacts of Network Congestion

In a wireless fault-tolerant consensus system, network congestion is one of the leading issues that influence system performance and stability. This part will delve into an in-depth analysis of the causes of network congestion and its implications on such systems.

Network congestion is usually caused by multiple factors. These include an abundance of data transmission requests, limited bandwidth resources, changes in network topology, and potential disturbances and conflicts. During scenarios where numerous nodes in the network are simultaneously sending and receiving data, the flood of data requests can monopolize substantial bandwidth resources. This causes other data transmission requests to be delayed or even dropped, thereby increasing transmission latency. Considering wireless networks have finite bandwidth resources, surpassing this capacity leads to data packet loss or delays in transmission. In severe cases, it could even bring about interruptions in the communication links. Frequent joining and leaving of mobile nodes in the network trigger changes in the network topology, necessitating frequent rerouting and adding to the network's load. This entails that these algorithms should not only exhibit high performance as the network size grows but also display resilience in the face of an increased number of potential faults [7]. This also destabilizes data transmission paths, increasing the possibility of transmission delays and data losses. Furthermore, wireless channels can be disrupted by other devices or environmental factors, leading to channel conflicts. These disruptions cause the channels to be occupied, preventing other data from timely transmission and triggering data errors that necessitate retransmission thus further increasing network load.

The primary manifestations of network congestion on wireless fault-tolerant consensus systems include first increased data transmission latency, which extends the consensus process timeframe. This leads to delays in system decision-making, impacting real-time response and potentially making prompt handling of sudden occurrences unfeasible. Second, network congestion leads to data packet loss during transmission, compromising the completeness and precision of consensus data. Lost data packets could cause inconsistencies between nodes, influencing the consensus result. To ensure data integrity, retransmission of data is necessary, further increasing network load and delay. Furthermore, severe network congestion could disrupt communication links, preventing nodes from participating in the consensus process normally. This interruption could cause the consensus process to fail or need reinitiation, and frequent communication interruptions hamper the overall system stability and reliability. Lastly, network congestion escalates competition for limited network resources (such as bandwidth and channels), possibly delaying or downgrading the consensus process and impacting consensus efficiency. Unequal distribution of resources might place certain nodes at a disadvantage in the consensus process, affecting system fairness.

Network congestion significantly impacts the performance, stability, and fairness of wireless fault-tolerant consensus systems. It increases consensus delay, leads to loss of data packets, causes communication interruptions, and hinders competitive resource allocation. Therefore, understanding the triggers and effects of network congestion and implementing effective congestion management systems are of utmost importance. These strategies can improve the efficiency and reliability of these systems. Future research should concentrate on creating smart network congestion prediction and prevention procedures and dynamically optimizing network resource allocation. This will ensure the steady and productive operation of wireless fault-tolerant consensus systems despite network congestion.

2.2 Fundamentals of Fault-Tolerant Wireless Consensus

2.2.1 Multicast in Wireless Consensus

This chapter examines multicast communication to groups of processes with known membership. The system under consideration consists of a set of processes that can communicate reliably over one-to-one channels, as shown in Fig. 2.1. As previously stated, processes may fail only by crashing.

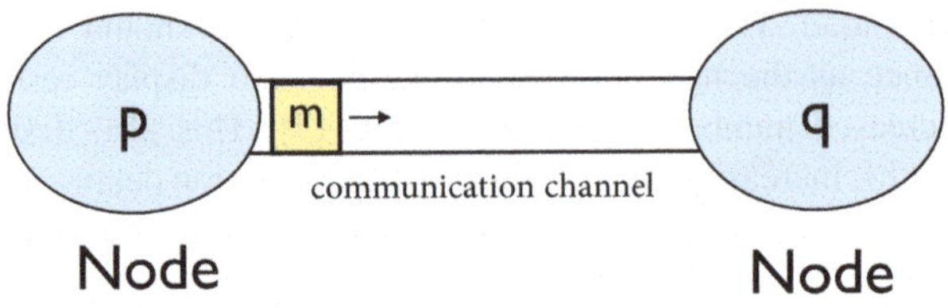

Fig. 2.1 Node communication model

Nodes are members of groups, which serve as the destinations for messages sent via the multicast operation. Allowing processes to simultaneously belong to multiple groups is generally beneficial, such as enabling processes to receive information from various sources by joining several groups. However, to simplify the discussion of ordering properties, processes may sometimes be restricted to being members of at most one group at a time.

The multicast (g, m) operation sends the message m to all members of the group g of processes. Correspondingly, the deliver (m) operation delivers a message sent by multicast to the calling process. The term "deliver" is used instead of "receive" to clarify that a multicast message is not always handed to the application layer inside the process immediately upon receipt at the process's node. This distinction is explained further when discussing multicast delivery semantics.

Every message m carries the unique identifier of the node sender (m) that sent it and the unique destination group identifier group (m). It is assumed that nodes do not misrepresent the origin or destinations of messages.

2.2.1.1 Fundamentals of Multicast

Having a basic multicast primitive that guarantees, unlike IP multicast, that a correct process will eventually deliver the message as long as the multicaster does not crash is useful. This primitive is called B-multicast, and its corresponding basic delivery primitive is B-deliver. Processes are allowed to belong to several groups, and each message is destined for a particular group.

A straightforward implementation of B-multicast using a reliable one-to-one send operation is as follows:

To B-multicast (g, m): for each process $p \in g$, send (p, m).

On receive (m) at p : B-deliver (m) at p.

The implementation may utilize threads to perform the send operations concurrently in an attempt to reduce the total time taken to deliver the message. However, such an implementation is prone to suffer from "ack-implosion" if the number of processes is large. The acknowledgments sent as part of the reliable send operation are likely to arrive from many processes at approximately the same time.

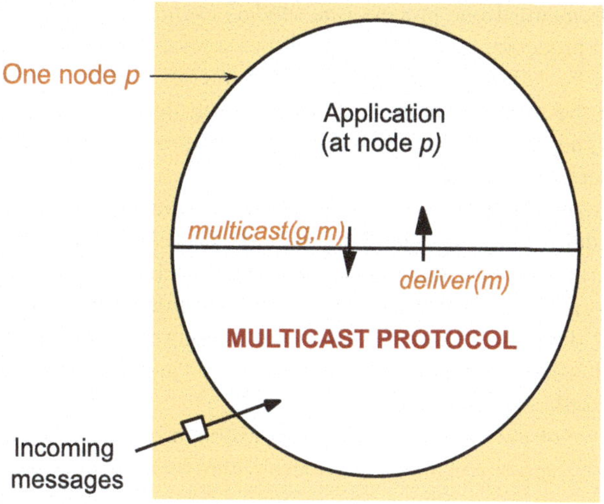

Fig. 2.2 Multicast model

2.2.1.2 Reliable Multicast

The former chapter discussed reliable multicast in terms of validity, integrity, and agreement. This section expands on that informal discussion by presenting a more complete definition.

Following Hadzilacos and Toueg [8] and Chandra and Toueg [9], a reliable multicast with corresponding operations R-multicast and R-deliver is defined. Properties analogous to integrity and validity are clearly highly desirable in reliable multicast delivery, but an additional requirement is added: all correct processes in the group must receive a message if any of them do. It is crucial to understand that this is not a property of the B-multicast algorithm based on a reliable one-to-one send operation. The sender may fail at any point during the B-multicast process, resulting in some processes delivering a message while others do not.

A reliable multicast shown in Fig. 2.2 satisfies the following properties:

Integrity: A correct process p delivers a message m at most once. Furthermore, $p \in$ group (m), and m was supplied to a multicast operation by sender (m). (As with one-to-one communication, messages can always be distinguished by a sequence number relative to their sender.)

Validity: If a correct process multicasts message m, then it will eventually deliver m.

Agreement: If a correct process delivers message m, then all other correct processes in group (m) will eventually deliver m.

The integrity property is analogous to that of reliable one-to-one communication. The validity property guarantees liveness for the sender. This property may seem unusual due to its asymmetry (mentioning only one particular process). However, it is important to note that validity and agreement together amount to an overall liveness requirement: if one process (the sender) eventually delivers a message m and the correct processes agree on the set of messages they deliver, it follows that m will eventually be delivered to all the group's correct members.

Expressing the validity condition in terms of self-delivery offers the advantage of simplicity. The requirement is that the message be eventually delivered by some correct member of the group.

The agreement condition is related to atomicity, the "all or nothing" property, applied to the delivery of messages to a group. If a process that multicasts a message crashes before delivering it, the message may not be delivered to any process in the group. However, if it is delivered to some correct process, then all other correct processes will deliver it. Many papers in the literature use the term "atomic" to include a total ordering condition. Many papers use "atomic multicast" to describe a kind of total ordering multicast.

The objective of the traditional consensus problem in a distributed system is to attain unanimous agreement among all nodes within the system concerning specific shared data values. Some prior works achieve consensus among all participating nodes by relying on a dependable wireless channel, e.g., [2] and [3]. In contrast, real-world wireless networks contend with an unreliable and error-prone shared wireless channel. The piece discusses a binary consensus issue in unreliable wireless networks involving n agents in a 3D Euclidean space. The agents, which can either agree or disagree with an event, can have any starting point and route. The only requirement is that they stay within each other's broadcast range. There's a maximum of f faulty agents, which can either crash or give false results. The non-faulty agents come to a consensus about the event. Essential principles have to be observed to reach a consensus despite faults in the distributed systems.

- Termination: Each non-faulty process reaches a decision. This implies that the consensus algorithm must guarantee that the consensus-reaching process ultimately ends, even in the presence of faulty nodes or an unreliable communication network.
- Agreement: Ultimately, all non-faulty nodes must converge on a single value or decision. This implies that the consensus algorithm should avoid scenarios where some nodes choose one value while others choose a different one.
- Liveness: The consensus algorithm is expected to make continuous progress in achieving consensus as long as the majority of nodes are correct, and the network maintains adequate connectivity. Liveness guarantees that the algorithm does not become stuck in a state where no further progress is possible.
- Validity: The accepted value must be one of the values originally proposed by a non-faulty node. In other words, the consensus algorithm should not permit nodes to reach a consensus on values that were not proposed by any non-faulty node.
- Fault Tolerance: The consensus algorithm should demonstrate resilience in the face of a defined number of faulty nodes or Byzantine faults. Even if a subset of nodes behaves unpredictably, the algorithm should maintain consensus among the non-faulty nodes.

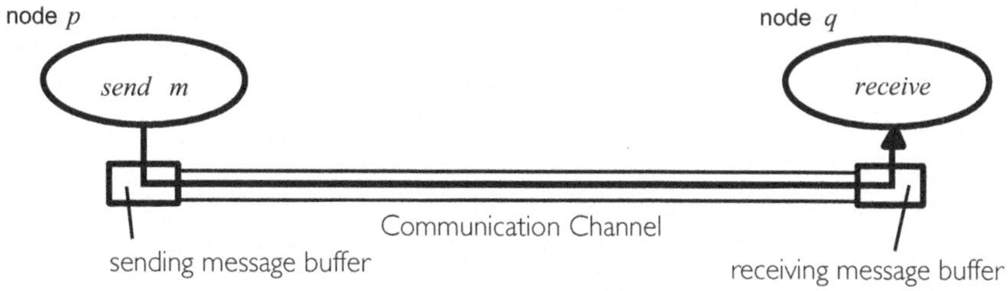

Fig. 2.3 Channel fault model

2.2.1.3 Multicast in Fault-Tolerant Wireless Networks

In the realm of fault-tolerant consensus, reliable communication and coordination among multiple processes are of paramount importance. The previous chapter delved into the concept of reliable multicast, a fundamental communication primitive that ensures the delivery of messages to a group of processes while satisfying key properties such as integrity, validity, and agreement. Reliable multicast forms the bedrock upon which many higher-level distributed protocols and applications are built, enabling efficient dissemination of information and facilitating collaboration among distributed entities.

We can get more detailed types of omission failure in Fig. 2.3

- Channel Omission. The message is omitted by the channel.
- Sending Omission. The process finalizes the "send" operation; however, the message does not reach its intended outgoing message buffer.
- Message Reception Failure. The message successfully arrives at the incoming message buffer; however, it is not received by the respective process.

However, as we shift our focus to wireless networks, a new set of challenges emerges that threatens to undermine the effectiveness and reliability of traditional multicast mechanisms. Wireless networks, with their unique characteristics and constraints, pose significant obstacles to the seamless operation of multicast protocols, necessitating a reevaluation of the assumptions and design principles that have proven successful in wired environments.

The advent of wireless networks has revolutionized the way we connect and communicate, offering unprecedented flexibility, mobility, and ubiquity. From mobile ad hoc networks (MANETs) to wireless sensor networks (WSNs), these networks have found applications in a wide range of domains, including military operations, disaster response, environmental monitoring, and smart cities. The ability to establish communication links without the need for fixed infrastructure has opened up new possibilities and enabled the deployment of distributed systems by fault-tolerant systems in previously unimaginable scenarios.

Yet, the very features that make wireless networks so appealing also introduce a host of challenges that complicate the design and implementation of reliable multicast protocols. The dynamic and unpredictable nature of wireless links, coupled with the mobility of nodes, creates a highly volatile environment where communication failures and network partitions are commonplace. The broadcast nature of the wireless medium, limited bandwidth resources, and energy constraints of battery-powered devices further compound the difficulties faced by multicast protocols.

In wired networks, reliable multicast protocols have been extensively studied and successfully deployed, leveraging the stability and reliability of point-to-point communication channels. Protocols such as Pragmatic General Multicast (PGM) and Reliable Multicast Transport Protocol (RMTP) have demonstrated their effectiveness in delivering messages to multiple recipients with strong consistency guarantees. These protocols often rely on the assumption of a fixed network topology, reliable links, and the availability of sufficient bandwidth and computational resources.

However, when transplanted to the wireless domain, these assumptions no longer hold true. The dynamic topology of wireless networks, where nodes can join, leave, or move at any time, renders the maintenance of stable multicast groups and delivery paths an arduous task. The inherent unreliability of wireless links, subject to interference, fading, and obstacles, makes it challenging to ensure the reliable delivery of messages to all intended recipients. The limited bandwidth and energy resources of wireless devices further constrain the design space, necessitating the development of protocols that are both efficient and parsimonious in their resource utilization.

Moreover, the failure of traditional multicast protocols in wireless networks has far-reaching implications that extend beyond the realm of communication itself. Many fault-tolerant consensus protocols, which are crucial for maintaining consistency and agreement among distributed processes, rely on the availability of reliable multicast primitives. Consensus protocols such as Paxos, RAFT, and Viewstamped Replication assume the existence of reliable communication channels and the ability to achieve atomic multicast, where all correct processes deliver the same set of messages in the same order.

The breakdown of these assumptions in wireless environments poses significant challenges to the design and implementation of fault-tolerant consensus protocols. The absence of reliable multicast undermines the foundation upon which these protocols are built, necessitating a rethinking of the approaches and mechanisms used to achieve consensus in the presence of failures and network partitions.

In this chapter, we embark on a thorough exploration of the challenges posed by multicast in wireless networks and their implications for fault-tolerant consensus protocols. We begin by examining the fundamental limitations of traditional multicast mechanisms in the face of the unique characteristics and constraints of wireless environments. Through rigorous analysis and formal proofs, we demonstrate the impossibility of guaranteeing the properties of atomic multicast in wireless networks with unreliable links and mobile nodes.

Building upon this foundation, we delve into the implications of the failure of atomic multicast for fault-tolerant consensus protocols. We highlight the need for novel approaches and design principles that take into account the specific challenges of wireless networks, such as handling unreliable communication, dealing with node mobility, ensuring consistency in partitioned networks, achieving energy efficiency, and supporting scalability.

Limitations of Multicast in Wireless Networks Wireless networks introduce a set of unique challenges that significantly impact the performance and reliability of traditional multicast mechanisms. These limitations stem from the inherent characteristics of wireless environments, which differ substantially from their wired counterparts. In this section, we briefly outline the key factors that contribute to the difficulties faced by multicast in wireless networks.

First, wireless networks are characterized by the mobility of nodes, which can freely move within the network coverage area. This mobility leads to frequent changes in the network topology, making it challenging to maintain stable multicast group memberships and delivery paths. The dynamic nature of wireless networks introduces inconsistencies in multicast message delivery, as nodes may move in and out of range, causing intermittent disruptions in connectivity.

Second, wireless links are inherently less reliable compared to wired links. Wireless communication is susceptible to various factors such as signal fading, interference, obstacles, and environmental conditions. These factors contribute to high packet loss rates, making it difficult to guarantee the reliable delivery of multicast messages to all intended recipients. The unreliability of wireless links poses significant challenges in ensuring the consistency and integrity of multicast communication.

Third, wireless networks typically have limited bandwidth resources compared to wired networks. The shared nature of the wireless medium leads to contention among multiple nodes, resulting in reduced available bandwidth for each node. This limited bandwidth constrains the efficiency and scalability of multicast communication, particularly when dealing with large amounts of data or a high number of recipients.

Lastly, energy constraints are a critical consideration in wireless networks, especially in scenarios involving battery-powered devices such as sensor nodes or mobile devices. The energy-intensive nature of multicast communication, which involves delivering messages to multiple nodes, can quickly drain the battery of these devices. The need for energy efficiency necessitates the design of multicast protocols that minimize energy consumption while still achieving the desired level of reliability and performance.

These limitations of multicast in wireless networks have been discussed in detail in the section "Causes and Impacts of Unstable Communication in Fault-tolerant Wireless Consensus." The challenges posed by node mobility, unreliable links, limited bandwidth, and energy constraints collectively contribute to the instability and unreliability of communication in wireless environments, which in turn hinders the effectiveness of traditional multicast mechanisms.

Recognizing and understanding these limitations are crucial for designing and implementing multicast protocols that can operate effectively in wireless networks. The following sections delve deeper into the specific implications of these limitations on fault-tolerant consensus protocols and explore potential approaches to overcome these challenges.

2.2.2 *Atomic Multicast in Wireless Networks*

Atomic multicast, also known as total order multicast, is a fundamental communication primitive that ensures all recipients receive messages in the same order. It is a crucial building block for many fault-tolerant distributed systems, including consensus protocols. However, the unique challenges posed by wireless networks make it exceptionally difficult to achieve atomic multicast in these environments. This section delves into the reasons behind the failure of atomic multicast in wireless networks and provides a comprehensive analysis supported by formal proofs and mathematical expressions.

- **System Model and Assumptions.** To set the stage for our analysis, we first define the system model and assumptions as follows: The wireless network consists of a set of nodes, denoted as $N = \{n_1, n_2, ..., n_k\}$. Nodes communicate through wireless links, which are subject to packet loss, delay, and corruption. Nodes can join, leave, or move within the network at any time, resulting in a dynamic topology. Each node has a unique identifier and maintains a local copy of delivered messages. The network may experience partitions, where subsets of nodes become disconnected from each other.
- **Properties of Atomic Multicast.** Atomic multicast aims to satisfy the following properties:
  - Validity: If a correct node multicasts a message m, then all correct nodes eventually deliver m.
  - Agreement: If a correct node delivers a message m, then all correct nodes eventually deliver m.
  - Integrity: For any message m, every correct node delivers m at most once, and only if m was previously multicast by a node.
  - Total Order: If correct nodes p and q both deliver messages m_1 and m_2, then p delivers m_1 before m_2 if and only if q delivers m_1 before m_2.
- **Challenges in Unreliable Wireless Networks.** Wireless networks introduce several challenges that hinder the realization of atomic multicast.
  - Unreliable Communication. Wireless links are inherently unreliable and prone to packet loss, delay, and corruption. Let P_L denote the probability of packet loss, P_D the probability of packet delay, and P_C the probability of packet corruption. The reliability of a wireless link can be expressed as:

$$R_{link} = (1 - P_L) \times (1 - P_D) \times (1 - P_C).$$

    As P_L, P_D, and P_C are nonzero in wireless networks, R_{link} is always less than 1, indicating that perfect reliability is unattainable.
  - Node Mobility. Nodes in wireless networks can move freely, leading to frequent changes in network topology. Let P_M denote the probability of a node moving out of the communication range of another node. The stability of a wireless link between two nodes, denoted as S_{link}, can be expressed as:

$$S_{link} = 1 - P_M.$$

As P_M is nonzero in mobile wireless networks, S_{link} is always less than 1, indicating that link stability cannot be guaranteed.

- Network Partitions. Wireless networks are susceptible to partitions, where subsets of nodes become disconnected from each other. Let P_P denote the probability of a network partition occurring. The connectivity of the network, denoted as C_{net}, can be expressed as:

$$C_{net} = 1 - P_P.$$

As P_P is nonzero in wireless networks, C_{net} is always less than 1, indicating that perfect network connectivity is impossible.

- **Impossibility of Atomic Multicast in Unreliable Wireless Networks.** In a wireless network with unreliable links, node mobility, and the possibility of network partitions, it is impossible to guarantee the properties of atomic multicast. We prove the impossibility of atomic multicast in wireless networks by contradiction. Assume that there exists an algorithm A that guarantees the properties of atomic multicast in a wireless network.

  Consider a scenario where a correct node n_i multicasts a message m. Due to unreliable communication, the probability of successful message delivery to all correct nodes is given by $P_{success} = R_{link}^N$, where N is the number of correct nodes in the network. As R_{link} is less than 1, $P_{success}$ approaches 0 as N increases, violating the validity property. Due to node mobility, the probability of maintaining stable links between all correct nodes during the multicast operation is given by $P_{stability} = S_{link}^{N(N-1)/2}$, where N is the number of correct nodes in the network. As S_{link} is less than 1, $P_{stability}$ approaches 0 as N increases, violating the agreement property. Due to the possibility of network partitions, the probability of maintaining a fully connected network during the multicast operation is given by $P_{connectivity} = C_{net}$. As C_{net} is less than 1, there is a nonzero probability of a network partition occurring, violating the total order property. Since the algorithm, A, violates the validity, agreement, and total order properties, it contradicts our assumption that it guarantees the properties of atomic multicast.

  Therefore, by contradiction, we conclude that it is impossible to guarantee the properties of atomic multicast in a wireless network with unreliable links, node mobility, and the possibility of network partitions.
- **Impact on Fault-Tolerant Consensus Protocols.** The impossibility of atomic multicast in wireless networks has significant implications for fault-tolerant consensus protocols that rely on it. Many consensus algorithms, such as Paxos, RAFT, and Chandra-Toueg, assume the availability of reliable communication channels and the ability to achieve atomic multicast. However, in wireless environments, these assumptions break down, rendering the protocols ineffective. Consider a consensus protocol that requires atomic multicast for message

ordering and agreement. In wireless networks, the protocol may encounter the following issues.

- Message Loss. Due to unreliable communication, messages may be lost, leading to inconsistent views among nodes and failure to reach consensus.
- Node Mobility. As nodes move in and out of communication range, the consensus protocol may struggle to maintain a consistent membership and ensure that all nodes receive the necessary messages.
- Network Partitions. In the presence of network partitions, the consensus protocol may reach different decisions in different partitions, violating the agreement property.

The impact of these issues on the correctness and performance of fault-tolerant consensus protocols can be severe. Let $P_{consensus}$ denote the probability of reaching a correct consensus decision in the presence of failures. In a wireless network, $P_{consensus}$ can be expressed as:

$$P_{consensus} = P_{success} \times P_{stability} \times P_{connectivity}.$$

As $P_{success}$, $P_{stability}$, and $P_{connectivity}$ are all less than 1 in wireless networks, $P_{consensus}$ is also less than 1, indicating that the reliability of consensus decisions is compromised.

- **Alternative Approaches for Wireless Fault-Tolerant Consensus.** Given the challenges and limitations of atomic multicast in wireless networks, researchers have explored alternative approaches to achieve consensus in these environments.

  - Gossip-Based Protocols. Gossip-based protocols rely on probabilistic message dissemination, where nodes exchange messages with a subset of their neighbors in each round. While these protocols do not provide strong consistency guarantees, they can be more resilient to message loss and node failures.
  - Hierarchical Consensus. Hierarchical consensus protocols partition the network into smaller groups or clusters, each with its own local consensus mechanism. The local decisions are then aggregated at higher levels to reach a global consensus. This approach can mitigate the impact of unreliable links and node mobility by localizing the consensus process.
  - Geographic Consensus. Geographic consensus protocols leverage the location information of nodes to make consensus decisions. By considering the geographic proximity of nodes, these protocols can minimize the impact of node mobility and network partitions.
  - Virtual Communication Protocols. Virtual consensus protocols combine multiple mechanisms, such as gossip-based dissemination and hierarchical consensus, to achieve a balance between consistency and reliability. These protocols adapt their behavior based on the network conditions and application requirements.

  While these alternative approaches offer potential solutions for achieving consensus in wireless networks, they often come with trade-offs in terms of performance, scalability, and complexity. The choice of an appropriate consensus mechanism depends on the specific characteristics and requirements of the wireless network and the application.

- **Case Study of Consensus in Wireless Sensor Networks**
  To illustrate the challenges of achieving consensus in wireless networks, let us consider a case study of wireless sensor networks (WSNs). WSNs consist of a large number of resource-constrained sensor nodes deployed in an area to monitor physical or environmental conditions. These nodes typically communicate through low-power, short-range wireless links and are subject to node failures, link failures, and network partitions.

  Suppose a WSN is tasked with reaching a consensus on the average temperature reading across all nodes. Each node measures the local temperature and needs to agree on the global average. In a traditional setting with reliable communication, this could be achieved using a simple averaging consensus protocol. However, in a wireless setting, the protocol faces several challenges:

  – Message Loss. Due to the unreliable nature of wireless links, temperature readings transmitted by nodes may be lost, leading to an incomplete or inaccurate global average.
  – Node Failures. Sensor nodes are prone to failures due to limited battery life, hardware malfunctions, or environmental factors. The failure of nodes can disrupt the consensus process and lead to inconsistent decisions.
  – Network Partitions. The deployment area of a WSN may be large, and nodes may be scattered across different regions. Network partitions can occur due to obstacles, interference, or node failures, preventing nodes from communicating with each other and reaching a consensus.

  To address these challenges, a WSN may employ a combination of techniques, such as:

  – Redundancy. Deploying multiple nodes in each region to provide redundancy and mitigate the impact of node failures
  – Multi-hop Communication. Implementing multi-hop communication protocols to enable nodes to relay messages across longer distances and bypass failed nodes or links
  – Localized Consensus. Partitioning the network into smaller clusters and running localized consensus protocols within each cluster, then aggregating the results at higher levels to obtain a global consensus

Despite these techniques, achieving perfect consensus in a wireless sensor network remains a challenging task. The inherent limitations of wireless communication and the resource constraints of sensor nodes impose fundamental trade-offs between accuracy, reliability, and efficiency.

2.2.3 *Atomic Multicast and Fault-Tolerant Consensus*

Traditional fault-tolerant consensus protocols, such as Paxos, RAFT, and Chandra-Toueg, have been widely used in distributed systems to ensure agreement among nodes in the presence of failures. These protocols rely heavily on the availability of reliable communication channels and the ability to achieve atomic multicast or other forms of reliable multicast. Atomic multicast guarantees that all nodes receive messages in the same order, providing a strong foundation for building consistent distributed systems.

In a typical fault-tolerant consensus protocol, nodes exchange messages to propose values, vote on proposals, and reach a common decision. Atomic multicast plays a crucial role in ensuring that all nodes receive the same sequence of messages, enabling them to make consistent decisions and maintain a coherent state. The properties of atomic multicast, such as validity, agreement, integrity, and total order, are essential for the correct functioning of these protocols.

For example, in the Paxos protocol, nodes assume the roles of proposers, acceptors, and learners. Proposers suggest values, acceptors vote on the proposed values, and learners learn the decided value. Atomic multicast is used to disseminate proposals and votes among the nodes, guaranteeing that all acceptors receive the messages in the same order. This ensures that the acceptors make consistent decisions and that the learners eventually learn the same decided value.

Similarly, in the RAFT protocol, nodes are organized into a leader-follower architecture. The leader is responsible for managing the replication of the log entries to the followers. Atomic multicast is used to ensure that all followers receive the log entries in the same order, maintaining a consistent state across the nodes. The leader election process also relies on atomic multicast to ensure that all nodes agree on the new leader.

The Chandra-Toueg consensus protocol, which is based on the concept of unreliable failure detectors, also depends on atomic multicast. Failure detectors provide information about the suspected failures of nodes, and the consensus protocol uses this information to make progress. Atomic multicast is employed to disseminate the failure detector information and the protocol messages, ensuring that all nodes have a consistent view of the system state.

The reliance on atomic multicast in these protocols is not limited to the dissemination of protocol messages. Many fault-tolerant consensus protocols also use atomic multicast to implement a replicated state machine. In a replicated state machine, each node maintains a copy of the application state, and the consensus protocol ensures that all nodes apply the same sequence of operations to their local state. Atomic multicast is used to propagate the operations to all nodes, guaranteeing that they are applied in the same order, resulting in a consistent state across the replicas.

The dependence on atomic multicast in fault-tolerant consensus protocols extends beyond the basic communication primitives. These protocols often rely on higher-level abstractions and mechanisms built on top of atomic multicast. For instance, group membership protocols, which manage the dynamic membership of nodes in a distributed system, often use atomic multicast to ensure that all nodes have a consistent view of the group membership. Failure detection and recovery mechanisms, such as checkpointing and state transfer, also depend on atomic multicast to ensure that all nodes have a consistent state after a failure.

The reliance on atomic multicast in traditional fault-tolerant consensus protocols is not without its challenges. As discussed in the previous section, achieving atomic multicast in wireless networks is extremely difficult, if not impossible, due to the unique characteristics of these environments. Unreliable communication, node mobility, and network partitions make it challenging to guarantee the strong consistency properties required by atomic multicast.

Consequently, the failure of atomic multicast in wireless networks has significant implications for the applicability and effectiveness of traditional fault-tolerant consensus protocols in these environments. Without the availability of reliable communication channels and the ability to achieve atomic multicast, these protocols may fail to provide the necessary consistency guarantees, leading to inconsistent decisions, divergent states, and system failures.

2.2.4 Designing Fault-Tolerant Consensus in Wireless Networks

The dependence on atomic multicast in traditional fault-tolerant consensus protocols, as discussed in the previous section, highlights the need for a fundamental redesign of consensus mechanisms for wireless networks. The unique characteristics and challenges of wireless environments render the existing approaches ineffective and necessitate the development of novel solutions that can cope with the inherent unreliability and dynamicity of these networks.

Wireless networks, such as mobile ad hoc networks (MANETs), wireless sensor networks (WSNs), and the Internet of Things (IoT), exhibit properties that are significantly different from those of traditional wired networks. These properties include high node mobility, unstable link quality, limited bandwidth, energy constraints, and network partitions. The failure of atomic multicast in wireless networks, as proven in the previous sections, underscores the impossibility of directly applying traditional fault-tolerant consensus protocols in these environments.

The reliance on strong consistency guarantees, such as those provided by atomic multicast, is no longer feasible in wireless networks. The assumptions of reliable communication channels and stable network topology, which are fundamental to the correctness and efficiency of traditional consensus protocols, do not hold in wireless settings. Consequently, attempting to employ these protocols in wireless networks may lead to poor performance, inconsistent decisions, and system failures. By redesigning fault-tolerant consensus protocols based on these considerations, researchers and practitioners can develop consensus mechanisms that are specifically tailored to the challenges and characteristics of wireless networks. These redesigned protocols will be better equipped to handle the unreliability, dynamicity, and resource constraints of wireless environments, enabling the deployment of reliable and efficient distributed systems in a wide range of wireless applications. The redesign of fault-tolerant consensus protocols for wireless networks is not a straightforward task and requires a deep understanding of the failure models and characteristics of these environments. The unique properties of wireless networks, such as high node mobility, unstable link quality, and resource limitations, introduce new dimensions of failures that need to be carefully considered in the design of consensus mechanisms.

In the following section, we delve into the specific failure models and characteristics of wireless networks, providing a comprehensive analysis of the challenges and opportunities for designing robust and efficient consensus protocols for these environments. By understanding the intricacies of wireless network failures, we can lay the foundation for the development of fault-tolerant consensus mechanisms that can effectively cope with the unreliability and dynamicity of wireless systems. The redesign of fault-tolerant consensus protocols for wireless networks represents a significant research challenge and an opportunity for innovation. It requires a multidisciplinary approach, combining insights from distributed systems, wireless communications, and fault tolerance. By leveraging the unique properties of wireless networks and exploring novel consensus mechanisms, researchers can unlock the potential of reliable and efficient distributed systems in a wide range of emerging wireless applications, such as IoT, vehicular networks, and drone swarms.

In conclusion, the dependence on atomic multicast in traditional fault-tolerant consensus protocols highlights the need for a fundamental redesign of consensus mechanisms for wireless networks. The failure of atomic multicast in these environments necessitates the development of alternative approaches that can cope with the inherent unreliability and dynamicity of wireless systems. By embracing the unique characteristics of wireless networks and considering the specific failure models and challenges, researchers can pave the way for the design and optimization of fault-tolerant consensus protocols that are specifically tailored to the needs of wireless distributed systems.

2.3 Failures in Fault-Tolerant Wireless Consensus

With the ongoing emergence of new types of wireless distributed applications such as the Internet of Things, vehicular networks, and drone swarms, the question of how to achieve reliable, efficient distributed consensus in a dynamic, unreliable wireless network environment has become a hotly debated research topic of mutual concern within both the academic and industrial spheres. Compared to traditional wired networks, wireless networks present the challenges of high node mobility, unstable link quality, and serious resource limitations. The situation thus demands higher requirements and greater challenges for distributed consensus. Existing consensus algorithms, such as Paxos and RAFT, are mostly designed for stable and reliable network environments, making them ill-suited for application within wireless scenarios. To break through this bottleneck, an in-depth understanding of the failure models and characteristics of wireless networks is necessary, upon which a design and optimization of fault-tolerant consensus protocols suitable for wireless environments can then be constructed.

Network failure models can be divided into two broad categories according to the behavioral characteristics of faulty nodes: non-Byzantine fault models and Byzantine fault models. Non-Byzantine fault models assume that nodes will only exhibit nonsubjective, unintentional error behaviors, such as crashes, delays, or occasional data errors, and will not launch malicious attacks. These error behaviors are usually caused by objective factors such as hardware failures, network interference, energy exhaustion, and so forth, making them relatively easy to predict and tolerate. Most existing distributed consensus algorithms, such as Paxos and RAFT, are designed based on non-Byzantine fault models. In contrast, Byzantine fault models further consider the possibility of malicious behavior by nodes, such as message forgery, selective forwarding, and conspiracy attacks. These necessitate the introduction of more complex security measures, such as encryption authentication and multi-signature mechanisms. Despite the relatively narrower application scope of non-Byzantine fault models, adopting these models remains a pragmatic choice for resource-limited wireless networks when considering the balance between security, performance, and costs. This section primarily focuses on non-Byzantine fault models and systematically analyzes the types, causes, and effects of failures in wireless networks from the perspectives of the physical layer, protocol layer, and application layer. It also explores the design considerations and challenges of fault-tolerant consensus protocols in wireless environments. Jing and Zou et al. conducted an examination of the detrimental impacts instigated by Byzantine nodes across the physical, protocol, and data layers. Their premise posited that these nodes possessed an awareness of the historical and current state of protocol execution in each round and retained the autonomy to behave arbitrarily within the physical, protocol, and data layers. Therefore, when categorizing consensus failures, these failures can be classified into the physical, protocol, and data layers in Fig. 2.4.

2.3.1 Physical-Layer Failures

Physical layer failures in non-Byzantine wireless networks mainly stem from the openness and shared nature of wireless channels. Compared to wired channels, wireless channels are more susceptible to environmental noise, signal attenuation, and multipath effects, which negatively impact signal quality and degrade communication reliability. Moreover, wireless channels represent a typically shared medium, with multiple nodes sharing limited spectral resources. When several nodes access the channel simultaneously, communication conflicts due to mutual interference can easily arise, preventing the receiving node from correctly demodulating signals.

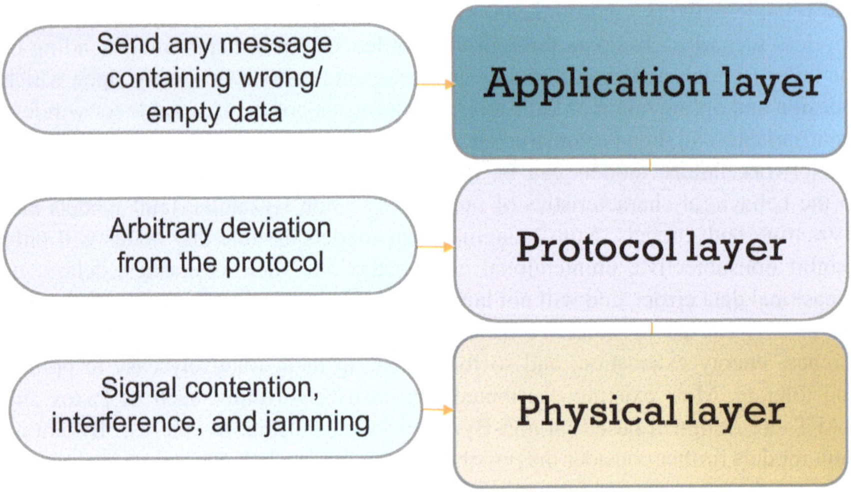

Fig. 2.4 Three-layer failure model

Communication conflicts are closely related to the transmitter and receiver characteristics of nodes. Due to limitations in costs and energy consumption, wireless nodes are usually equipped with simple single-antenna half-duplex transceivers. Under half-duplex mode, a node cannot receive signals while sending them, requiring time division multiplexing of the channel for sending and receiving. The single-antenna architecture further restricts the spatial resolution capacity of the node. Overall, only when the signal strength received by the node is sufficiently high can the target signal be separated from interference and noise, realizing reliable communication. Once multiple nodes simultaneously transmit signals to the receiving node, the node has difficulty accurately decoding information from each sending node due to antenna gain constraints, significantly reducing communication quality. If the number of nodes sending simultaneously further increases, the signal-to-interference-and-noise ratio at the receiving node will fall below the decoding threshold, leading to total communication disruption.

Communication conflicts are also related to network topology and traffic distribution. In distributed networks like wireless ad-hoc networks, the deployment of nodes is usually random and dynamically changing. Compared with regular, static network topology, random topology more easily results in a dense distribution of nodes in local areas, aggravating communication conflicts. Even under relatively uniform node distribution, dynamic changes in business traffic could trigger local communication conflicts. For example, in event-driven sensor networks, when an event occurs within a monitored area, the sensor nodes in that area typically initiate data collection and reporting simultaneously, forming a surge in business traffic. This causes a sudden increase in local communication loads and intensifies channel contention. Therefore, when designing fault-tolerant consensus protocols, the impact of network topology and traffic distribution on physical layer communication reliability should be fully considered.

Besides channel contention, wireless communication is also readily subject to external disturbances. First, wireless radio waves attenuate as the distance increases during transmission. Under a free-space channel, the energy of electromagnetic waves attenuates at a rate proportional to the square of the distance. When the communication distance surpasses a certain range, the signal strength at the receiving node will fall below the receiver's sensitivity, rendering signal detection and decoding ineffective. Second, wireless radio waves can be obstructed and reflected by obstacles such as terrain and buildings. When a communication link is blocked by an obstacle, the signal strength rapidly diminishes, severely deteriorating communication quality. The unevenness of the obstacle's surface can also result in multipath propagation effects, causing signal distortion. Lastly, environmental noise and other electronic devices can significantly interfere with wireless communication. The frequency band of wireless channels is usually not specifically allocated, requiring sharing with other wireless systems. When the electromagnetic radiation from other systems enters the communication channel, it acts as additional noise, lowering the signal-to-noise ratio. At the same time, the radio frequency circuit of the communication device itself can also generate harmonic interference, affecting the signal reception of nearby devices. The resistance of wireless communication to interference is closely related to modulation methods and transmission power. Generally speaking, modulation methods with higher spectral efficiency tend to have weaker anti-interference capabilities, and lower transmission power has a lower tolerance for interference. Therefore, when designing fault-tolerant consensus protocols, the communication ability and anti-interference ability of wireless nodes must be balanced, necessitating the selection of appropriate physical layer configurations.

Energy restrictions represent another important factor affecting the reliability of the physical layer in wireless networks. Compared to nodes in wired networks, wireless nodes typically rely on batteries or energy-harvesting devices for power, consequently limiting available energy. Communication forms the major energy consumption source for wireless nodes. When transmitting or receiving data, wireless transceivers need to provide a certain transmission power or reception sensitivity, with energy consumption far exceeding that of data processing and storage. When nodes engage in communication over a long duration, the battery rapidly drains, thereby shortening the node's lifespan. Although nodes would not maliciously consume energy under non-Byzantine fault models, depletion of energy can still cause nodes to temporarily or permanently fail, prompting them to leave the network. As more and more nodes malfunction, network connectivity gradually deteriorates, making it difficult to ensure the reliability and viability of the consensus process. Also, the node density in some areas will drop significantly, with reduced channel competition and lower spectrum utilization. Therefore, when designing fault-tolerant consensus protocols, the communication load of individual nodes should be minimized as much as possible to extend network life expectancy. Implementing mechanisms such as node rotation and hibernation to balance energy consumption levels among nodes, thereby avoiding premature failure of individual nodes, offers an effective way to improve the network's overall robustness. Such failures can result in unforeseen disruptions in the system's operation, ultimately detrimentally affecting performance and reliability. For instance, in scenarios involving crash failures or transient failures, Bhusalmith et al. [10] demonstrated that non-faulty nodes could experience temporary disruptions, preventing them from reaching a consensus. Furthermore, in the context of accidental failures within wireless sensor networks, Liu et al. [11] pointed out that unreliable wireless communication can lead to message loss and node failures.

2.3.2 Protocol Layer Failures

Protocol layer failures form another prominent factor impacting the reliability of wireless networks. Compared to physical layer failures, protocol layer failures are more often attributable to software implementation flaws and anomalies incited by network dynamic changes. First, as a type of distributed software, the implementation of a consensus protocol may inherently possess unavoidable defects, such as algorithm bugs, memory leaks, deadlocks, and so on. When a defect occurs in the protocol implementation of a certain node, it may lead to state errors, message loss, non-responsiveness, and other abnormal behaviors, in turn influencing the protocol execution of other nodes and the final consensus result. Second, even with perfectly implemented protocols, unanticipated events stemming from network dynamic changes could inadvertently push nodes to deviate from the protocol. For instance, when a node temporarily loses connection, it may miss key consensus messages, causing its state to diverge from that of other nodes. When a node moves, its connection relationship with other nodes might change, rendering the previously existing communication mode unsuitable and leading to message delay, disorder, or loss. The dynamical joining and leaving of nodes could also impact the consensus process and alter the trust basis for decision-making. For example, in blockchain technology, extensively utilized in cryptocurrencies and decentralized applications, consensus protocols such as PoW and PoS are utilized to guarantee agreement among network nodes regarding transaction order and validity. Furthermore, the work of RAFT 184040 addressed the consensus challenge in blockchains by employing non-Byzantine fault-tolerant algorithms.

At the protocol layer, for example in the CSMA/CA protocol, a data packet is transmitted after the distributed inter-frame spacing (DIFS) time and the backoff time, which is uniformly distributed in the range $[0, CW - 1]$. CW is the value of the contention window. For the first backoff, CW is initialized as CW_{min}. In the 802.11 standard,

$$CW = 16, 32, 64, 128, 256, 512, 1024,$$

and $CW_{\text{min}} = 16$, $CW_{\text{max}} = 1024$.

$$\overline{\text{BackoffTime}} = (1-p)\frac{CW_{\min}}{2} + \ldots + p^m(1-p)\frac{\sum_{i=0}^{m} 2^i CW_{\min}}{2} +$$
$$\ldots + p^{m_r}\frac{\sum_{i=0}^{m} 2^i CW_{\min} + (m_r - m)\,2^m CW_{\min}}{2}.$$

The bit error rate (BER) is a measure of the accuracy of data transmission within a specified period of time.

$$\text{Bit error rate} = \frac{\text{Bit errors in transmission}}{\text{The total number of transmitted codes}} \times 100\%.$$

If there are bit errors, there is a bit error rate. In addition, the BER is defined as the frequency at which bit errors occur. The IEEE802.3 standard defines a maximum acceptable BER of 10^{-10}. It is important to note that the unreliable communication provided at the physical layer is a major inducement of errors at the protocol layer. Due to the random fluctuations in wireless link quality, supposedly reliable message exchanges may fail to be completed within the anticipated timeframe, leading to protocol anomalies. For example, when a node broadcasts a proposal, the degradation of individual link quality could cause certain nodes to fail to receive the proposal on time, or the confirmation message feedback could fail to reach it on time, triggering inconsistencies in protocol states. When a node requests a consensus proof, the loss of individual proof messages may result in the requesting node being unable to gather enough proof within the expected timeframe, leaving it stuck in perpetual waiting. In short, communication anomalies at the physical layer further magnify the uncertainties during the execution of the protocol, escalate state divergence, and break the consistency of consensus outcomes. Therefore, when designing fault-tolerant consensus protocols, the robustness of protocol execution needs to be strengthened on the basis of dealing with the unreliability of the physical layer.

2.3.3 Data Layer Failures

Unlike the physical layer, non-Byzantine failures at the data or information layer highlight that participating nodes may inadvertently transmit messages with incorrect, erroneous, or empty content. This manipulation aims to mislead honest nodes in the quest for attaining a prevailing consensus. For example, in distributed database systems, consensus mechanisms can be employed to guarantee that updates, inserts, and deletions in the database receive consensus agreement from all relevant nodes. This serves to mitigate problems like data disparities or conflicts that might emerge due to network problems or node failures, such as fake messages. In their work [12], Koponen et al. addressed non-Byzantine failures in distributed databases. When non-Byzantine nodes experience crashes or network partitioning, achieving consensus becomes an insurmountable challenge.

Data layer failures represent the third type of factor impacting the reliability of wireless fault-tolerant consensus. Unlike the physical and protocol layers, the application layer mainly focuses on the actual services provided by the network, such as perception, computation, control, and so on. Data layer failures typically manifest as errors or discrepancies in service outcomes, like biases in environmental status estimation, inconsistencies in computational results, etc. Even though the impact of application layer failures on consensus is relatively indirect, these failures can skew consensus results and affect decision reliability due to their role as an important input into the consensus process. For example, in a consensus-based distributed learning setting, consensus nodes usually have to agree on global model parameters. Owing to the random error of wireless channels, raw sample data gathered from the physical environment may be biased. When these biases are significant, they can lead to nodes fitting to different local models, consequently affecting the accuracy of the global model. In distributed situational awareness, the network needs to form a uniform understanding of the global situation. Due to differences in perceptual range, nodes in varying regions may perceive disparate local features. Simple data fusion might not eliminate these discrepancies and, therefore, affect the network's estimation of situational results.

State discrepancies are another significant source of data layer failures. Due to the distributed nature of wireless networks, there could be disparities in different nodes' environmental state perceptions. For instance, a node located in the center of interest might detect an event occurrence sooner, while nodes at the periphery might be unable to observe the same event due to limited perceptual range. Network partitioning can further intensify these state discrepancies. When a network undergoes segmentation due to node mobility or link interruptions, any separated areas will evolve distinct network states and are unable to exchange up-to-date information with each other. Message delay and loss could also create distortions in nodes' global state awareness. Based on this incomplete or outdated state information, nodes might infer incorrect conclusions, affecting the final consensus result. The influence of application layer failures on consensus is indirect, yet more covert and harder to address.

In non-Byzantine wireless networks, the physical, protocol, and application layers are all vulnerable to a variety of failure factors. These factors interact with one another, collectively influencing the reliability and stability of networks. These failures present serious challenges to fault-tolerant consensus in wireless networks. Traditional consensus algorithms are ill-suited for direct application within wireless scenarios, necessitating novel improvements and optimizations specifically for the unique properties of wireless networks.

2.3.4 Methods of Handling Failures

In the scope of this chapter, a "single layer" denotes a specific layer within the network stack where fault-tolerant wireless consensus (FTWC) is implemented. The network stack, commonly referred to as the networking model, typically comprises multiple layers, including the physical, data link, network, transport, and application layers. Each layer serves a distinct role in the process of transmitting data across the network. In a single-layer FTWC approach, fault-tolerant mechanisms are implemented exclusively at one specific layer of this stack [4].

Numerous studies have investigated various fault models in the domain of distributed computing, with the primary focus on solving the fault-tolerant consensus problem at a single layer. These studies encompass a wide range of subfields, methodologies, and solutions. Starting at the physical layer, faults mainly manifest as accidental failures, such as hardware malfunctions, power loss, or unintentional misconfigurations. These failures can lead to unexpected disruptions in the system's operation, adversely impacting performance and reliability. Bhusalmith et al. [10] made significant contributions to understanding these issues. Their algorithm was designed to handle crash failures, ensuring that non-faulty nodes reach agreement even in the presence of unexpected failures. They also extended their research to address transient failures, which may be temporary but can still cause significant disruptions. Similarly, in wireless sensor networks, Liu et al. [11] specifically considered accidental failures. They tackled the unreliable nature of wireless communication by introducing an algorithm that efficiently manages message loss and non-malicious node failures. This study highlighted the often unpredictable nature of wireless networks, where even minor disturbances can result in widespread communication failures. Furthermore, in vehicular networks, Lee et al. [13] investigated achieving consensus in the presence of transient faults caused by environmental factors. They proposed a robust consensus algorithm that accounts for dynamic network topology and vehicle mobility, addressing the unique challenges of real-time communication in constantly changing conditions. They also explored the potential of using multiple communication channels to reduce the likelihood of failure due to environmental interference.

Extending the discussion to protocol layers, in the context of distributed databases, Koponen et al.'s work [12] plays a foundational role in managing non-Byzantine failures. They developed protocols to ensure that a system can reach a consensus and continue operating even in the face of node crashes or network partitioning. Their work laid the groundwork for subsequent studies that aimed to enhance robustness and scalability in various distributed computing environments. Zhang et al. [14] delved into fault-tolerant consensus algorithms resilient to non-malicious failures such as link failures and node crashes at the protocol layers. They developed an adaptive consensus protocol capable of dynamically adjusting to changing network conditions, offering an effective solution tailored to the unique challenges of distributed computing, thereby enhancing system resilience.

At the data layer, Leslie Lamport's work in his paper "The part-time parliament," published in the ACM Transactions on Computer Systems [15], introduced the Paxos consistency algorithm based on message passing. This algorithm primarily addresses how a distributed system can reach consensus on a specific value. As distributed consensus research evolved, the Paxos algorithm gained widespread acceptance in both academic and industrial circles. Thompson et al. [16] addressed software bugs and unintentional errors within cloud computing environments. They employed a replication technique to handle non-malicious faults, ensuring continuous functionality even when some replicas encounter unexpected errors. This approach is critical in cloud environments, where redundancy and replication are key to maintaining availability and reliability. They further examined the trade-offs between performance and fault tolerance, emphasizing the need to strike a balance that aligns with the specific requirements of different applications and services. Furthermore, in the realm of consensus algorithms for blockchains, these algorithms are generally non-Byzantine fault-tolerant, such as RAFT, widely used in distributed databases and predominantly in consortia, chains, and private chains like ETCD [17] and BRAFT [18]. These algorithms also need to account for non-malicious failures. Although the original Bitcoin consensus algorithm was primarily designed to withstand Byzantine failures, subsequent research has focused on optimizing consensus in the presence of accidental faults such as network delays or node crashes. Adaptation and innovation in this field have led to a plethora of consensus algorithms explicitly designed to function in the inherently decentralized and distributed nature of blockchain technologies.

The challenge of achieving non-Byzantine fault tolerance in a single layer encompasses a wide spectrum of research across various fields, spanning from the design of physical hardware to application levels, and from working with wireless sensors to developing blockchain technologies. The objective is to create systems capable of withstanding accidental and non-malicious breakdowns. The referenced research covers a diverse array of strategies and solutions, each tailored for specific scenarios and assumptions about fault characteristics and the need for fault tolerance. The ongoing growth and diversification of this research underscore the constantly evolving nature of technology and computing, necessitating ongoing adaptation and innovation. The inherent complexity in these systems calls for a comprehensive and nuanced approach, involving cross-disciplinary collaboration and the integration of diverse methods and perspectives. The collective knowledge and achievements in this field form a solid foundation for continuous exploration and refinement of fault-tolerant distributed computing systems, underscoring the importance and relevance of this critical research area.

2.4 Cross-Layer Optimization

Cross-layer fault-tolerant algorithms play a crucial role in ensuring the reliability and robustness of fault-tolerant wireless consensus systems. These algorithms leverage the interaction and collaboration between different layers of the network stack to enhance the system's ability to tolerate faults and maintain consensus among nodes. In this section, we will first focus on the abstract MAC (absMAC) layer approach, which provides a fault-tolerant communication abstraction for higher-layer protocols. We will discuss the implementation of a fault-tolerant absMAC layer and how it can be utilized to build fault-tolerant wireless consensus protocols.

Furthermore, we will explore other types of cross-layer fault-tolerant algorithms that employ different strategies and techniques to achieve fault tolerance in wireless consensus systems. By understanding and leveraging these cross-layer approaches, we can design more resilient and efficient fault-tolerant wireless consensus systems that can operate reliably in the presence of faults and failures.

2.4.1 Abstract MAC Layer

The abstract MAC purpose is to provide a solid foundation for the theoretical study of higher-level algorithms in a MANET environment. This abstract service layer allows algorithm designers to avoid addressing low-level issues such as contention management and collision detection, instead encapsulating their effects through abstract time delay boundaries. The MAC layer is a critical component in the Open Systems Interconnection (OSI) model and is responsible for managing access to the shared communication medium in a network. In wireless networks, the challenges are particularly pronounced due to the inherent nature of the wireless medium. The MAC layer is the second layer in the OSI model, sitting just above the physical layer. Its primary role is to determine how devices in a network gain access to the shared communication medium, whether it is a wireless channel or a wired connection.

The concept of an abstract MAC layer refers to a higher-level, generalized representation of the MAC layer's functionalities. It abstracts and generalizes the core principles and mechanisms used for medium access control, making it easier to understand and design protocols that operate at this layer, as shown in Fig. 2.5. In wireless networks, the abstract MAC layer deals with unique challenges. These challenges include contention for the shared wireless medium, dealing with interference, and ensuring efficient utilization of the limited available bandwidth. The abstract MAC layer provides a framework for addressing these challenges. The abstract MAC layer is not limited to wireless networks. In distributed computing, especially in scenarios where multiple nodes or processors need to access shared resources, a similar concept can be applied. Here, it is about managing access to shared resources efficiently, avoiding conflicts, and ensuring fair access.

The abstract MAC layer serves as a foundation for designing specific MAC protocols and algorithms. For example, in wireless networks, the IEEE 802.11

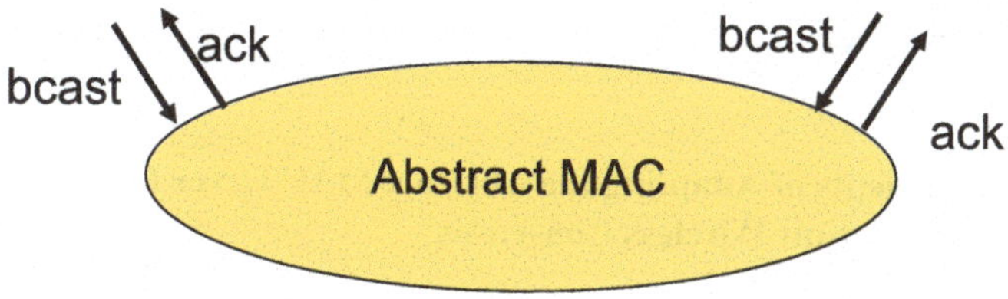

Fig. 2.5 Abstract MAC layer model

standard defines multiple MAC protocols (e.g., CSMA/CA for Wi-Fi) [19], each of which is based on the abstract MAC layer principles. The key functions associated with the abstract MAC layer include channel access control, collision avoidance (to prevent multiple devices from transmitting simultaneously), and fairness in resource allocation. These functions are implemented differently in various MAC protocols, depending on the network's specific requirements. In both wireless networks and distributed computing environments, the abstract MAC layer plays a crucial role in optimizing the performance of the network or distributed system. This optimization involves trade-offs between efficiency, fairness, and the ability to handle varying traffic loads.

2.4.1.1 The Main Advantages of the Abstract MAC Layer

The abstract MAC layer has the following advantages:

- It reliably delivers messages within the local vicinity, providing the sender with feedback that the message has been successfully sent to all nearby receivers.
- It provides guaranteed upper bounds for the worst-case time for message delivery to all receivers, as well as the total time for the sender to receive confirmation. These time assurances are represented by a time delay function that monitors the current contention levels within the sender and receiver zones.
- It encapsulates the details of the contention management mechanism within the service implementation, thus providing a simple abstract model for the algorithm designers that only involves assurances of message delivery and time boundaries.
- It provides a simple and realistic foundation for the study of higher-level algorithms and lower bounds within MANET. The algorithm designers can use these abstract time delay functions to analyze the algorithms regardless of specific channel behaviors.
- By implementing the abstract MAC layer on a basic radio network model, specific definitions for these time delay functions can be generated. This automatically applies the time delay boundaries proven for the abstract service to the particular radio network under consideration.

Therefore, the abstract MAC layer, by providing a reliable local broadcast along with contention-related time delay boundaries, simplifies the design and analysis of fault-tolerant algorithms in complex wireless environments. It effectively encapsulates the lower-level details, allowing designers to focus on the higher-level algorithms themselves.

2.4.1.2 The Benefits of Adapting the Abstract MAC Layer for Fault-Tolerant Wireless Consensus

The abstract MAC layer model significantly simplifies the design and analysis of fault-tolerant consensus algorithms in wireless networks by providing a suitable abstraction that captures the unique characteristics and challenges of the wireless medium. The key aspects of the model that facilitate the development of fault-tolerant wireless consensus solutions are given as follows:

Broadcast Communication In the abstract MAC layer model, nodes communicate by broadcasting messages rather than using point-to-point channels. This communication style naturally fits the characteristics of wireless networks and does not require nodes to have prior knowledge of other participants' identities. It simplifies the design of fault-tolerant consensus algorithms, as they no longer need to rely on a priori network knowledge, avoiding complex network discovery and management.

Limited Network Knowledge Unlike classic asynchronous message-passing models, nodes in the abstract MAC layer model do not need to know the network size or participant information. While this may seem like a limitation, it actually encourages algorithms to adopt techniques that are independent of network size, such as the counter-based approach mentioned in the chapter. This makes the algorithms more robust and adaptable, capable of working in networks of various scales.

Unique Asymmetry Due to the broadcast nature of wireless communication, the abstract MAC layer model exhibits a fundamental asymmetry: the sender of a message does not know which specific nodes have received the message, while the receivers know the source of the message. This asymmetry leads to a certain degree of asynchrony, making classic synchronous algorithms inapplicable. However, the sender receives an acknowledgment (ACK) after the message has been delivered to all nodes, providing a valuable synchronization point. Cleverly, leveraging this asymmetry and limited synchrony can greatly simplify the design and correctness proofs of fault-tolerant consensus algorithms.

Reasonable Fault Model The chapter considers the crash fault model, where nodes may permanently stop during execution. The abstract MAC layer model provides a reasonable abstraction of crash faults: if a sender crashes during the broadcast process, it may lead to only a subset of nodes receiving the message. Algorithms must be able to handle such inconsistent situations. Notably, due to the presence of ACKs, a node either receives the complete message or not at all, eliminating certain complex fault scenarios and simplifying fault-tolerant reasoning. Moreover, the assumption that crashed nodes do not recover further reduces the burden on algorithms.

Moderate Asynchrony Despite the presence of limited synchrony (e.g., ACKs), the abstract MAC layer model remains an asynchronous model. The FLP impossibility [1], as proven by Fischer, Lynch, and Paterson, states that it is impossible to solve consensus in asynchronous message-passing systems prone to even a single process crash failure. However, the model makes some realistic assumptions about message delivery (e.g., crashes only occur during the broadcast process) and introduces ACKs as synchronization points, leaving room for algorithms to overcome the impossibility. The chapter leverages these characteristics of the model to design randomized fault-tolerant consensus algorithms, proving that fault-tolerant consensus is solvable in this model.

Alignment with Wireless Environments The abstract MAC layer model abstracts typical characteristics of wireless networks, such as broadcast communication, collision management, and unstable link quality, making it more suitable for wireless scenarios compared to asynchronous message-passing models. Consequently, fault-tolerant consensus algorithms designed in this model are more likely to be efficiently implemented in real wireless systems. Algorithms can take advantage of the features provided by the model (e.g., ACKs) to avoid unnecessary complexity.

The abstract MAC layer model simplifies the complexity of the fault-tolerant consensus problem by appropriately abstracting the wireless environment while retaining the key characteristics of wireless networks. The model's unique properties, such as broadcast communication, limited network knowledge, asymmetry, reasonable fault model, and moderate asynchrony, provide favorable conditions for designing efficient fault-tolerant consensus algorithms. These features alleviate the burden on algorithms and open up new design possibilities, making problems that are difficult to solve in the asynchronous message-passing model become solvable in the abstract MAC layer model. This highlights the significant impact of computational model choice on the solvability and complexity of distributed algorithms.

2.4.2 Fault-Tolerant Abstract MAC Layer

For a successful execution and application of a robust, fault-tolerant abstract MAC layer, which is fundamentally a vital building block in a network protocol, there are several crucial factors that need to be taken into account and thoroughly examined. It is wholeheartedly integral for anyone looking to carry out this implementation that their approach carefully encompasses and thoroughly examines the following key elements or characteristics.

- Employ existing mature wireless Medium Access Control (MAC) layer technologies to manage channel contention, such as Carrier Sense Multiple Access (CSMA), Time Division Multiple Access (TDMA), or randomized backoff algorithms (e.g., DECAY). These mechanisms can ensure that communication between other nodes in the network remains unaffected even when nodes crash or fail. In the case of CSMA, an acknowledgment (ACK) is generated after the MAC layer detects that the channel is idle; in the case of TDMA, the ACK is generated after the node's own time slot ends; if using a randomized algorithm like DECAY, the ACK is generated after a sufficient number of retries to guarantee delivery to all neighboring nodes with high probability.
- Provide necessary reliability guarantees at the MAC layer, using mechanisms like retransmission and error correction to combat wireless channel interference and message loss, thus achieving the reliable broadcast semantics required by

the upper layer. If a broadcast is submitted to the abstract MAC layer, it will eventually be delivered to all neighboring nodes.

- Once the MAC layer determines that a message has been successfully sent (through its contention management and reliable transmission mechanisms), it returns an ACK to the upper layer protocol as confirmation. This ACK only notifies the upper layer that the message has been delivered and does not contain any information about the number or identity of the receiving nodes. The ACK contains no information about the number or identity of the message recipients.
- In the face of node crashes, the abstract MAC layer should ensure that other nodes remain unaffected and can continue to operate. The abstract MAC layer model assumed in the chapter can tolerate an arbitrary number of node crashes.
- When implementing the abstract MAC layer, no a priori knowledge about the number or identity of participating nodes in the network should be assumed. Our algorithms work for networks of any size and do not require advanced knowledge of the network participants.
- The interface provided by the abstract MAC layer to the upper layer protocol mainly includes two types of events: local broadcast and an ACK for each broadcast. Through this simple interface, the upper layer protocol can access the services provided by the abstract MAC layer.

The abstract MAC layer can be implemented using currently available wireless MAC technologies, such as CSMA, TDMA, and DECAY, by introducing necessary fault tolerance and reliability mechanisms at the MAC layer. At the same time, this abstract MAC layer model simplifies the design of upper-layer protocols by shielding many of the complex details of low-level wireless communication, providing a good foundation for designing efficient distributed algorithms.

In wireless networks or distributed computing, the abstract MAC layer model refers to a conceptual framework that defines how devices share access to the communication medium, such as the wireless spectrum or shared network resources. The MAC layer plays a crucial role in coordinating access to the medium, managing collisions, and ensuring efficient data transmission among multiple devices.

The abstract MAC layer model typically includes the following key components:

- **Channel Access Methods.** This component defines the mechanisms by which devices contend for access to the communication medium. Common channel access methods include CSMA, TDMA, and CDMA.
- **Medium Access Control Protocols.** MAC protocols govern the rules and procedures for accessing the communication medium. These protocols determine how devices contend for access, resolve collisions, allocate resources, and manage contention in a distributed manner.
- **Frame Format.** The MAC layer defines the structure and format of data frames transmitted over the network. This includes specifications for frame headers, payload, and error detection mechanisms, which enable devices to interpret and process data exchanged on the network.

- **Quality of Service (QoS) Management.** The MAC layer may incorporate mechanisms for managing QoS requirements, such as prioritizing traffic based on application-specific criteria, ensuring bandwidth guarantees, and minimizing latency and packet loss.

Overall, the abstract MAC layer model provides a high-level framework for understanding and designing MAC protocols and algorithms in wireless networks and distributed computing systems. It abstracts the complexities of the underlying hardware and physical layer technologies, allowing for the development of interoperable and efficient communication solutions across diverse network environments.

2.4.3 *Fault-Tolerant Wireless Consensus via Abstract MAC Layer*

Fault-tolerant consensus is a fundamental problem in distributed computing, and it has been extensively studied in the context of wireless networks. Currently, there are two state-of-the-art approaches to achieving fault-tolerant wireless consensus. Since this chapter focuses on non-Byzantine failures, we will consider the approach proposed by Newport and Robinson [20], which assumes a fault-tolerant abstract MAC layer. This work presents two randomized algorithms that achieve fault-tolerant consensus in wireless networks without requiring prior knowledge of the network size or participant information, making them suitable for networks of various scales. In the following, we will discuss these algorithms in detail:

In Counter Race Consensus, each node maintains a proposal value and a counter. Nodes continuously broadcast their proposal value and counter value, updating their state upon receiving an ACK. If a received proposal value has a higher counter than the node's own, the node adopts that proposal value and updates its counter to the received counter value plus 1. When a node's counter exceeds a threshold (which is $2n^3\lceil\log_2 n\rceil + 4n^2$ in the chapter), it decides on that proposal value and terminates the algorithm. The algorithm leverages counters to "weigh" proposal values. As the algorithm progresses, proposal values that are broadcast earlier and more frequently will have increasingly larger counters, eventually dominating. Conflicting proposal values may prevent some nodes' counters from reaching the threshold, causing them to remain undecided. The chapter introduces randomized backoff times in broadcasts to break such deadlocks, ensuring that proposals with larger counter-values eventually prevail. This algorithm achieves consensus with high probability within $O(n^3 \log n)$ broadcast rounds.

In Almost-Everywhere Consensus, the algorithm consists of two phases: the first phase estimates the network size n, and the second phase achieves consensus. In the first phase, each node generates a random bit string of length $\Theta(\log n)$ and sets it as its initial proposal value, propagating it through broadcasts. The number of collected proposal values can be used to estimate the network size. After

$\Theta(n \log^2 n)$ broadcast rounds, all nodes can obtain a constant factor estimate $\tilde{n}$ of n. In the second phase, nodes run multiple rounds of Counter Race Consensus, with each round aiming to eliminate some conflicting proposal values. After $O(\log n)$ rounds, it is guaranteed that a vast majority ($1 - O(1/\log n)$ fraction) of nodes reach consensus. The total number of communication rounds for both phases is $O(n \log^2 n)$. This algorithm achieves fast convergence while relaxing the consensus requirement, allowing a small fraction of nodes not to participate in the consensus.

Both algorithms leverage the ACK-based synchronous communication semantics provided by the abstract MAC layer to design two new randomized algorithms that achieve fault-tolerant consensus in wireless networks.

2.4.4 Other Methods of Cross-Layer Optimizaiton

In the continually evolving realm of network systems and distributed computing, the study of fault-tolerant wireless consensus (FTWC) across multiple layers emerges as a fundamental and intricate discipline. While traditional approaches often focus on a single layer, the concept of multilayer fault-tolerant consensus has gained traction in recent years. Unlike the three-layer approaches mentioned that fully support the physical layer, protocol layer, and application layer, the multilayer approach discussed in this chapter adopts a more flexible and adaptable perspective.

Within the context of this chapter, "multiple layers" denotes an approach in which FTWC mechanisms are integrated across various layers of the network stack, but not necessarily encompassing all three layers comprehensively. This approach draws inspiration from the cross-layer design philosophy, aiming to address various fault types at the most suitable layer while maintaining a balance between complexity and effectiveness [21]. By selectively applying fault-tolerant techniques across different layers, this multilayer approach enables a more targeted and efficient solution compared to the traditional three-layer approach.

The multilayer approach to FTWC offers a more holistic and adaptable comprehension of fault-tolerant consensus, allowing for the development of tailored solutions that cater to the specific requirements and constraints of different network environments. By leveraging the strengths of cross-layer design, this approach facilitates the creation of robust and resilient distributed systems capable of operating in the face of various faults and adversities.

At the foundational level of the physical layer, where it interfaces with the protocol layer, NBFTC has been employed to tackle a wide range of hardware-related issues. Faults at these levels often materialize as hardware malfunctions, signal loss, and link failures [22]. Research in this domain encompasses the creation of redundant communication routes and mechanisms for detecting and rectifying errors directly within the physical layer. This guarantees that even in the event of device failure, the data layer can persist in transmitting and receiving information through the protocol paths.

The combination of the protocol and application layers introduces a distinct set of challenges and possibilities when implementing FTWC. In this context, the focus shifts from pure hardware failures to matters concerning routing, congestion, and packet management [23, 24]. The protocol layer's responsibility for ensuring dependable data delivery across the network necessitates the use of dynamic and adaptive algorithms capable of withstanding transient failures. Techniques like adaptive routing and congestion control mechanisms designed to respond to non-malicious failures play a vital role within these layers. Research in this area often centers on ensuring the successful routing of data by application algorithms [25, 26], even in the presence of network failures or unexpected alterations in network topology.

The intersection between the physical and application layers offers a distinct vantage point on NBFTC. Here, the emphasis shifts toward ensuring data consistency across distributed systems while considering both software and hardware failures. Diverse consensus algorithms, such as Paxos [15] and RAFT [27], have been devised to address non-malicious failures within the abstract MAC layer [28]. The abstract MAC layer is a valuable tool for streamlining consensus design, ensuring that physical failures do not disrupt the application layer. These algorithms function by overseeing data replicas across different nodes [20], ensuring that even if a segment of the system encounters issues, overall functionality remains uninterrupted [29, 30]. Challenges at this level often revolve around preserving consistency and performance, given the heightened complexity and greater susceptibility to failures in a distributed environment.

2.5 Applications of Fault-Tolerant Wireless Consensus

Scenario I UAV Swarm Systems In the application scenarios of UAV swarms, UAVs can be used for tasks such as disaster rescue, environmental monitoring, and military reconnaissance. Through fault-tolerant wireless consensus technology, UAVs can achieve task coordination and decision-making consistency without a central controller. For example, in disaster rescue, a UAV swarm can negotiate the task allocation for each UAV through consensus algorithms, such as the division of search areas, to ensure coverage of the entire disaster area. In addition, through consensus algorithms, UAVs can collaboratively plan flight paths, avoid collisions with each other, and maximize coverage of the search area. During the execution of tasks, UAVs share information in real time through wireless communication and dynamically adjust their respective tasks and flight positions through consensus algorithms to ensure the efficiency and precision of rescue operations. If the UAV swarm cannot reach a consensus, it may lead to uneven task allocation, conflicting path planning, and even UAV collisions. This not only results in mission failure and resource waste but may also delay rescue timing, causing greater casualties and property losses.

Scenario II Smart Grid In the application scenarios of smart grids, the production, distribution, and consumption of electricity need to be monitored and dynamically adjusted in real time to ensure the stability and efficiency of the power supply. Fault-tolerant wireless consensus technology plays a key role here by helping multiple distributed energy nodes achieve coordination and optimization. Specifically, each power node in the smart grid can negotiate load allocation in real time through consensus algorithms to prevent the overloading of certain nodes and ensure system stability. At the same time, by sharing status information of each part through consensus algorithms, the system can detect and isolate faults in a timely manner, ensuring the reliable operation of the grid. In addition, distributed power generation equipment, such as solar panels and wind turbines, can also coordinate power generation and energy storage through consensus algorithms to optimize overall power supply and meet user demands. If the nodes in the smart grid fail to reach a consensus, it may lead to unbalanced power distribution, causing overloading or insufficient power supply in some areas, which in turn may cause grid failures that cannot be discovered and handled in time. This situation may trigger large-scale power outages, seriously affecting the normal operation of society and the economy. In summary, fault-tolerant wireless consensus technology is a core technology in smart grids to ensure system stability and efficiency and can significantly improve the reliability of power supply.

Scenario III Smart Agriculture In the application scenarios of smart agriculture, multiple sensors and devices are used to monitor and manage the growth conditions of crops to improve the efficiency and quality of agricultural production. Fault-tolerant wireless consensus technology is crucial in this field because it can ensure data consistency and coordinated management among various devices. For example, through consensus algorithms, sensors for soil moisture, temperature, and light can share and confirm environmental data, thereby optimizing irrigation and fertilization strategies to keep the crop growth environment in an optimal state. In addition, crop monitoring sensors can collaboratively detect pests and diseases through consensus algorithms, promptly discover problems, and take preventive measures to prevent the spread of pests and diseases and protect crop health. During the harvest season, agricultural machinery can also work collaboratively through consensus algorithms to optimize harvesting paths and times, improve harvesting efficiency, and reduce grain loss. If the devices in the smart agriculture system cannot reach a consensus, it may lead to inaccurate environmental data, which in turn affects irrigation and fertilization decisions, resulting in poor crop growth. At the same time, if pests and diseases are not detected and handled in a timely manner, it may cause large-scale crop damage or even total crop failure. In addition, uncoordinated harvesting may lead to collisions or repetitive operations of mechanical equipment, reducing harvesting efficiency and increasing operating costs. Overall, fault-tolerant wireless consensus technology ensures the consistency and coordination of data and operations in smart agriculture, greatly improving the efficiency and reliability of agricultural production.

2.6 Summary

Wireless consensus algorithms find applications in various distributed systems and wireless networks, including blockchain and cryptocurrency systems, distributed databases, cloud computing, and Internet of Things (IoT) networks. The breadth of applications highlights the universality of the consensus problem. It is a fundamental issue in distributed systems, and various industries and technologies rely on consensus algorithms to ensure reliability and agreement.

However, achieving consensus in wireless networks presents unique challenges compared to traditional wired networks. These challenges include variability in communication conditions (e.g., signal strength, interference, and latency), dynamic network topologies, and the potential for node mobility. Such factors can complicate the process of reaching an agreement among nodes in a wireless setting.

Different wireless consensus algorithms have trade-offs in terms of performance, fault tolerance, and scalability. The choice of algorithm depends on the specific requirements of the system and the desired trade-offs. The choice of wireless consensus algorithm is not one size fits all. It depends on the particular use case. Some algorithms might prioritize fault tolerance over performance, while others might focus on scalability. Understanding these trade-offs is essential when designing or selecting a consensus algorithm.

Fault-tolerant consensus algorithms typically have properties like safety and liveness. Safety ensures that nodes never decide on conflicting values, even in the presence of faults. Liveness guarantees that nodes eventually make a decision.

The choice of a consensus algorithm for a wireless network depends on factors like the specific requirements of the application, desired trade-offs between fault tolerance and performance, and the types of failures the network is likely to encounter. Some algorithms, like PBFT, are suitable for highly fault-tolerant scenarios, while others, like RAFT, prioritize simplicity. Fault-tolerant consensus algorithms find applications in various domains, including sensor networks, autonomous vehicles, IoT devices, and any scenario where distributed nodes need to agree on a common value or decision, even in the presence of unreliable communication. Researchers and engineers continuously work on improving and developing new consensus algorithms tailored to the specific challenges of wireless networks. They aim to optimize performance, reduce overhead, and enhance the fault tolerance capabilities of these algorithms.

This section explores the critical challenges posed by unstable communication in wireless fault-tolerant consensus systems. Real-world wireless networks are often unreliable and error-prone due to factors such as signal interference, node movement, device failures, insufficient signal coverage, and network congestion. These issues can lead to data inconsistencies, increased latency, resource waste, and weakened robustness, severely impacting the performance and coordination capability of the system. The section introduces atomic multicast as a communication method in traditional fault-tolerant consensus and highlights the inability of traditional fault-tolerant consensus algorithms to execute reliable multicast in unstable wireless networks as the primary hindrance to their effectiveness. To address this challenge, the abstract MAC layer protocol is introduced as a potential solution to enable reliable multicast and improve the performance of fault-tolerant consensus algorithms in unstable wireless networks.

References

1. Fischer, Michael J., Nancy A. L., & Michael S. P. (1985). Impossibility of distributed consensus with one faulty process. *Journal of the ACM (JACM), 32*(2), 374–382.
2. Yang, L., Zou, Y., Xu, M., Xu, Y., Yu, D., & Cheng, X. (2022). Distributed consensus for blockchains in internet-of-things networks. *TUP Tsinghua Science and Technology, 27*(5), 817–831.
3. Xu, Q., Zou, Y., Yu, D., Xu, M., Shen, S., & Li, F. (2020). Consensus in wireless blockchain system. In *Springer WASA* (Vol. 12384, pp. 568–579).
4. Barborak, M., Dahbura, A., & Malek, M. (1993). The consensus problem in fault-tolerant computing. *ACM Computing Surveys, 25*(2), 171–220.
5. Wu, J., & Dai, F. (2004). Mobility-sensitive topology control in mobile ad hoc networks. In *18th International Parallel and Distributed Processing Symposium, 2004. Proceedings* (p. 28).
6. Li, T., Fu, M., Xie, L., & Zhang, J.-F. (2010). Distributed consensus with limited communication data rate. IEEE Trans. Autom. Control **56**(2), 279–292 (2011)
7. Kirti, S., Scaglione, A., & Thomas, R. J. (2007). A scalable wireless communication architecture for average consensus (pp 32–37).
8. Hadzilacos, V., & Toueg, S. (1994). A modular approach to fault-tolerant broadcasts and related problems. Technical report, USA, 1994
9. Chandra, T. D., & Toueg, S. (1996). Unreliable failure detectors for reliable distributed systems. *Journal of the ACM, 43*(2), 225–267.
10. Bhusal, N., Abdelmalak, M., Kamruzzaman, M., & Benidris, M. (2020). Power system resilience: Current practices, challenges, and future directions. *IEEE Access, 8*, 18064–18086.
11. Liu, K., Abu-Ghazaleh, N., & Kang, K.-D. (2007). Location verification and trust management for resilient geographic routing. *Journal of Parallel and Distributed Computing, 67*(2), 215–228.
12. Koponen, T., Casado, M., Gude, N., Stribling, J., Poutievski, L., Zhu, M., Ramanathan, R., Iwata, Y., Inoue, H., Hama, T., & Shenker, S. (2010). Onix: A distributed control platform for large-scale production networks. In *9th USENIX Symposium on Operating Systems Design and Implementation (OSDI 10)*, Vancouver. USENIX Association.
13. Santini, S., Salvi, A., Valente, A. S., Pescapè, A., Segata, M., & Cigno, R. L. (2018). Platooning maneuvers in vehicular networks: A distributed and consensus-based approach. *IEEE Transactions on Intelligent Vehicles, 4*(1), 59–72.
14. Thambidurai, P., & Park, Y. K. (1988). Interactive consistency with multiple failure modes. In *Proceedings Seventh Symposium on Reliable Distributed Systems* (pp. 93–100). IEEE Computer Society.
15. Lamport, L. (1998) The part-time parliament. *ACM Transactions on Computer Systems, 16*(2), 133–169.
16. Zhou, A., Wang, S., Cheng, B., Zheng, Z., Yang, F., Chang, R.N., Lyu, M.R., & Buyya, R. (2016). Cloud service reliability enhancement via virtual machine placement optimization. *IEEE Transactions on Services Computing, 10*(6), 902–913.
17. Shi, F., Romano, A., Philips, B., Betz, J., Lee, G., Hu, J., Li, X., Darnell, B., Batschelet, S. Etcd. https://github.com/etcd-io/etcd
18. Lin, T. Braft. https://github.com/baidu/braft
19. IEEE 802.11. (1999). Wireless lan mac and physical layer specifications.

20. Newport, C., & Robinson, P. (2018). Fault-tolerant consensus with an abstract MAC layer. In U. Schmid & J. Widder (Eds.), *DISC* (Vol. 121, pp. 38:1–38:20).
21. Xu, T., & Potkonjak, M. (2016). Energy-efficient fault tolerance approach for internet of things applications. In *2016 IEEE/ACM International Conference on Computer-Aided Design (ICCAD)* (pp. 1–8).
22. Geeta, D.D., Nalini, N., & Biradar, R. C. (2013). Fault tolerance in wireless sensor network using hand-off and dynamic power adjustment approach. *Journal of Network and Computer Applications, 36*(4), 1174–1185.
23. Kim, J. H., Liu, Z., & Chien, A. A. (1997). Compressionless routing: a framework for adaptive and fault-tolerant routing. *IEEE Transactions on Parallel and Distributed Systems, 8*(3), 229–244.
24. Dimopoulos, M., Gang, Y., Benabdenbi, M., Anghel, L., Zergainoh, N. E., & Nicolaidis, M. (2013). Fault-tolerant adaptive routing under permanent and temporary failures for many-core systems-on-chip. In *2013 IEEE 19th International On-Line Testing Symposium (IOLTS)* (pp. 7–12).
25. He, T., Stankovic, J. A., Lu, C., & Abdelzaher, T. (2003). Speed: A stateless protocol for real-time communication in sensor networks. In *23rd International Conference on Distributed Computing Systems, 2003. Proceedings* (pp. 46–55).
26. He, T., Stankovic, J. A., Abdelzaher, T. F., & Lu, C. (2005). A spatiotemporal communication protocol for wireless sensor networks.
27. Ongaro, D., & Ousterhout, J. (2014). In search of an understandable consensus algorithm. In *2014 USENIX Annual Technical Conference (USENIX ATC 14)* (pp. 305–319). USENIX Association.
28. Kuhn, F., Lynch, N., & Newport, C. (2009). The abstract MAC layer. In I. Keidar (Ed.), *Distributed computing* (pp. 48–62).
29. Newport, C. (2014). Consensus with an abstract MAC layer. In *PODC* (pp. 66–75).
30. Yu, D., Zou, Y., Yu, J., Zhang, Y., Li, F., Cheng, X., Dressler, F., & Lau, F. C. M. (2020). Implementing the abstract MAC layer in dynamic networks. *IEEE Transactions on Mobile Computing, 20*(5), 1832–1845.

Chapter 3
Byzantine Fault-Tolerant Wireless Consensus

Byzantine faults describe the abnormal behavior of certain nodes in wireless systems due to hardware failures, network issues, software errors, or malicious attacks, which may be difficult to predict or understand by other normal nodes. Therefore, Byzantine fault-tolerant wireless consensus emerged. Byzantine fault-tolerant (BFT) consensus is a consensus mechanism aimed at addressing possible hardware errors, network congestion or interruptions, and malicious attacks in distributed systems. It ensures that nodes can reach consistent decisions in untrusted environments by solving Byzantine fault problems.

The Byzantine fault-tolerant wireless consensus has broad application scenarios, and a large number of studies have introduced Byzantine fault-tolerant wireless consensus algorithms tailored to the unique characteristics of specific scenarios, including the Internet of Things (IoT), in-vehicle networks, and drone networks. For example, in the field of the Internet of Things (IoT), [1] introduces a position-based scalable consensus protocol known as geographic practical Byzantine fault tolerance (GPBFT). Recognizing that most IoT systems rely on stationary devices for data collection and processing, GPBFT leverages the geographic information of these fixed-position devices with greater computational capabilities to achieve consensus. In this protocol, nodes with relatively stable positions are selected as endorsers in the practical Byzantine fault tolerance (PBFT) consensus process, thereby reducing the overhead associated with transaction validation and recording. In another approach, [2] proposes a scalable multilayer PBFT consensus mechanism aimed at minimizing the communication volume of traditional single-layer PBFT consensus. This mechanism involves an analysis of the optimal node count at each layer to achieve minimal communication volume. Furthermore, [3] introduces an improved Byzantine consensus mechanism built on K-medoids clustering. In this mechanism, nodes participating in consensus are clustered based on specific features, and the algorithm is applied to the resulting hierarchical model. This adaptation significantly reduces the frequency of consensus-related communications, leading to enhanced consensus efficiency.

M. Xu et al., *Wireless Consensus*, Wireless Networks,
https://doi.org/10.1007/978-3-031-70859-6_3

In the field of vehicular networks, in [4], researchers introduce an innovative endorser selection mechanism that leverages reinforcement learning techniques. They also propose Byzantine fault-tolerant consensus protocols that are enhanced with reinforcement learning, ensuring that these consensus mechanisms can dynamically adapt to changing network conditions, varying vehicle behaviors, and malicious activities. This approach provides a resilient and intelligent framework for vehicular communication. In [5], a distributed fault-tolerant consensus algorithm is employed to achieve coordination among nearby vehicles. The proposed solution not only tolerates Byzantine faults but also accounts for lossy communication, offering a practical approach for vehicular networks. This ensures reliable coordination and data exchange among vehicles in challenging environments. Differing from existing research, [6] focuses on applying consensus-based fault-tolerant platooning for connected and autonomous vehicles. The objective is to detect faulty position and speed information originating from vehicles within a platoon and reinstate the normal behavior of the platoon.

In the field of drone networks, a lightweight asynchronous provable Byzantine fault-tolerant (LAP-BFT) consensus method is introduced in [7]. This method aims to reduce communication overhead by adopting a novel technique, which involves splitting the local trusted state transaction set and then decentralized reliable broadcast control transmission. In [8], a distributed MAC switching methodology leveraging Q-learning was outlined, encompassing MAC preselection evaluations and a reliable decision-making framework grounded in practical Byzantine fault tolerance (PBFT) to facilitate MAC protocol switching. The MAC preselection process allows individual drones to accurately assess their performance and identify the optimal MAC protocol for their current operational context. In [9], delved into the escalating research challenge posed by swarm drones and introduced an implementation strategy for Byzantine fault tolerance in swarm robots. Their approach is theoretically deemed more efficient and adaptable than previous methods, offering a framework that can be readily extended to analogous multi-agent collaborative tasks. This was achieved by integrating multi-agent foraging algorithms with Byzantine fault-tolerant (BFT) consensus algorithms, particularly for unmanned aerial vehicle group detection missions.

This chapter will introduce the positioning of the Byzantine consensus, common attack methods in the Byzantine fault-tolerant consensus, and corresponding defense methods in the Byzantine consensus. First, the definition and components of the Byzantine fault-tolerant system were introduced, with a focus on the core position of the Byzantine fault-tolerant consensus mechanism in the Byzantine fault-tolerant system. Next, common attack methods in Byzantine fault-tolerant consensus were introduced from four levels: physical layer, protocol layer, data layer, and cross-layer, such as data tampering, sybil attacks, and false reports. Then, in response to the above attack methods, corresponding defense methods in Byzantine fault-tolerant consensus were introduced. Existing research has continuously improved the defense methods of Byzantine fault-tolerant consensus from single-layer to multilayer. Finally, this chapter summarizes the Byzantine fault-tolerant wireless consensus and systematically elaborates on the above content.

3.1 Byzantine Fault Tolerance

The Byzantine fault-tolerant system is a mechanism that can ensure security and consistency in distributed systems. Its name originates from the problem of command transmission in the ancient Byzantine empire [10], which is described as follows:

As shown in Fig. 3.1, the Byzantine empire wanted to attack a powerful enemy and sent ten armies to surround it. This enemy, although not inferior to the Byzantine empire, is still capable of withstanding simultaneous attacks from five conventional Byzantine armies. For some reason, these ten armies cannot gather together for a single breakthrough and must attack simultaneously in a separate encirclement state. They have no chance of winning if any army attacks alone unless at least six armies attack simultaneously to capture the enemy country. They are scattered around the enemy country, relying on communication between signalmen to negotiate attack intentions and timing. The problem that troubles these generals is that they are not sure if there are traitors among them, and the traitors may change their attack intentions or timing without authorization. In this state, could the Byzantine generals find a distributed protocol that allowed them to negotiate and reach consensus remotely, thus winning the battle?

In distributed systems, nodes (which can be physically separated computers or logically separated processes) need to communicate and coordinate with each other to ensure the normal operation of the system. However, when there are communication failures or malicious nodes among the nodes, the system may exhibit unpredictable behavior. The Byzantine fault-tolerant system is designed to address this issue. In the Byzantine fault-tolerant system, each node communicates with each other, even if some of them may malfunction or be controlled by malicious nodes. This mechanism ensures that all honest nodes can reach a consistent state, maintaining the correctness and reliability of the system even in environments with potential failures and malicious behavior.

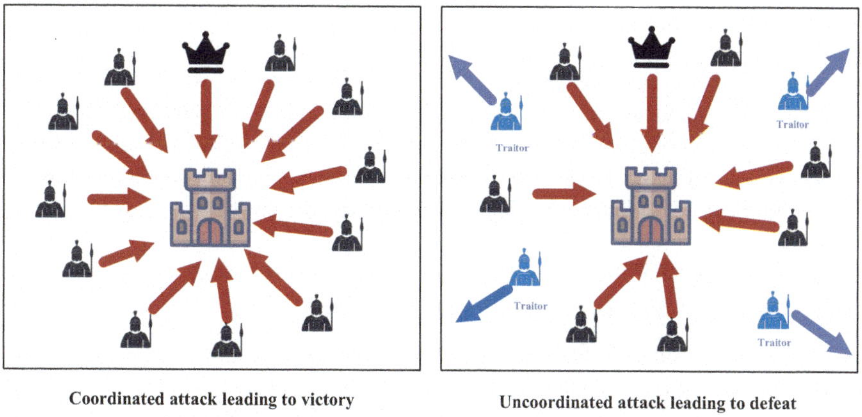

Fig. 3.1 The Byzantine generals problem

Specifically, the Byzantine fault-tolerant system encompasses multiple aspects such as network communication, consensus mechanisms, fault detection and isolation, fault-tolerant data replication, and performance optimization. Network communication ensures the integrity, authenticity, and confidentiality of messages during transmission; consensus mechanisms enable all normal nodes to reach agreement; fault detection and isolation can identify and isolate malicious or faulty nodes; fault-tolerant data replication guarantees the reliability and consistency of data; and performance optimization ensures efficient operation of the system in practical applications. The detailed description is given in the following.

I. Network communication involves the transmission and reception of messages, ensuring the integrity, authenticity, and confidentiality of data. It covers the following:

- Encryption technology. Encryption technology is a crucial component in data security. It utilizes various encryption algorithms to transform readable data into an encrypted format that is unreadable without the proper decryption key. This process ensures the confidentiality of sensitive data during transmission over networks, protecting it from unauthorized access or disclosure.
- Identity verification. Ensuring the authenticity of messages and their senders is crucial in secure communication. Identity verification techniques, such as digital signatures and certificates, provide a means to validate the identity of the sender. These techniques rely on cryptographic methods to attach a unique digital signature to a message, which can be verified by the receiver using the sender's public key. This ensures that the message has not been tampered with and is indeed from the intended sender.
- Message integrity verification. Message integrity verification is a key aspect of data security. Techniques like hash functions are used to create a unique fingerprint or hash value for a message. This hash value is then transmitted alongside the message. At the receiving end, the receiver calculates the hash value of the received message and compares it with the transmitted hash value. If they match, it indicates that the message has not been tampered with during transmission.
- Communication protocol. A communication protocol defines the rules and standards that govern the exchange of information between devices or systems. It specifies the message format, transmission method, error handling, and other critical aspects of communication. By adhering to a well-defined communication protocol, systems can ensure the reliability, consistency, and interoperability of communication, reducing the risk of data loss or corruption.

II. Consensus mechanism is the key to ensuring that all normal nodes can reach consensus on a certain state or value. It covers the following:

- Election Algorithm. In distributed systems and blockchain networks, an election algorithm is crucial for selecting leading or representative nodes. These nodes are responsible for initiating and coordinating the consensus process, which ensures that all nodes in the network agree on the current state of the ledger or database. The election algorithm ensures that these leading nodes are selected fairly and based on predefined criteria, such as their availability, performance, or stake in the network.

- Voting Mechanism. The voting mechanism in a consensus protocol allows nodes in the network to vote on proposed blocks or states. Each node casts its vote based on its validation of the proposed data. The consensus is reached when a sufficient number of nodes agree on a particular block or state, according to the rules defined in the consensus protocol. This voting mechanism ensures that malicious or faulty nodes cannot unilaterally influence the consensus outcome.
- Fault Tolerance. A key characteristic of robust consensus protocols is their ability to tolerate a certain number of malicious or faulty nodes. Fault tolerance mechanisms ensure that the system can still function properly and reach consensus, even when some nodes behave unexpectedly or maliciously. This is achieved through techniques such as Byzantine fault tolerance, which allows the system to identify and exclude misbehaving nodes from the consensus process.
- Performance Optimization. In order to improve the scalability and efficiency of consensus protocols, performance optimization techniques are employed. By optimizing algorithms and protocols, the latency and overhead in the consensus process are reduced, resulting in faster block confirmation times and improved system throughput. Performance optimization techniques can include optimizing the voting mechanism, reducing the number of messages exchanged between nodes, or utilizing parallel processing and caching techniques.

III. Fault detection and isolation are important means to ensure system robustness and reliability. It covers the following:

- Monitoring Mechanism. An effective monitoring mechanism is essential for maintaining the integrity and stability of distributed systems. It involves real-time monitoring of node behavior and status, allowing administrators or automated systems to detect anomalies or faults in a timely manner. By continuously collecting and analyzing data from nodes, the monitoring mechanism can identify potential issues and initiate corrective actions to prevent system failures.
- Diagnostic Techniques. When anomalies or faults are detected, diagnostic techniques are used to analyze the behavior logs, performance indicators, and other relevant data of the nodes. These techniques help identify the root cause of the faults, whether it is a software bug, hardware failure, or malicious activity. By understanding the cause, administrators can develop appropriate strategies to resolve the issue and prevent it from recurring.
- Isolation Strategy. Once malicious or faulty nodes are detected, an isolation strategy is implemented to separate them from the rest of the system. This ensures that the faulty nodes do not have any further impact on the operation of normal nodes and prevents the spread of any potential issues. Isolation can be achieved through techniques such as network segmentation, firewall rules, or by blocking communication with the faulty nodes.

- Recovery Mechanism. After the faulty node is isolated, a recovery mechanism is triggered to restore the normal operation of the system. This includes repairing or replacing the faulty hardware, updating or patching the software, or reconfiguring the system to mitigate the impact of the fault. The recovery mechanism ensures that the system can quickly return to a stable state and continue providing services to users.

IV. Fault-tolerant data replication is the key to ensuring data reliability and consistency. It covers the following:

- Data Backup. Implementing a robust data backup strategy is crucial for ensuring business continuity. Storing critical data in multiple replicas across geographically dispersed locations safeguards against data loss due to a single point of failure. This redundancy ensures that in case of any hardware failure, natural disaster, or other unforeseen event, data remains accessible and recoverable.
- Data Synchronization. Maintaining consistency and real-time data synchronization between all replicas is essential for maintaining data integrity. Utilizing technologies such as distributed consensus algorithms or replication protocols, data updates made to one replica are propagated to all others, ensuring that all replicas contain identical and up-to-date information. This synchronization ensures that users accessing the data from any replica receive the same accurate data.
- Data Recovery. In the event of a system failure, data loss, or corruption, having a reliable data recovery plan is vital. Through the use of backup copies, administrators can quickly restore data to its previous state, minimizing downtime and ensuring the continuity of business operations. The data recovery process should be tested regularly to ensure its effectiveness and readiness in case of an actual emergency.
- Load Balancing. Implementing an effective load-balancing strategy is crucial for optimizing system performance and scalability. Reasonably allocating the storage and access of data replicas based on system load and performance requirements ensures that no single replica becomes overloaded, thus maintaining optimal response times and throughput. Load balancing techniques such as round-robin, weighted distribution, or dynamic scaling can be employed to achieve this goal.

V. Performance optimization is the key to improving efficiency and reducing latency and overhead. It covers the following:

- Algorithm Optimization. By constantly refining and optimizing algorithms and data structures, we can significantly enhance the computational efficiency and accuracy of our systems. This process involves analyzing the current implementation, identifying bottlenecks and inefficiencies, and then implementing improvements to enhance the performance of critical operations. Algorithm optimization not only improves system speed but also reduces the likelihood of errors and increases the reliability of results.

- Protocol Optimization. The optimization of communication protocols and consensus mechanisms is crucial for reducing network latency and overhead. By evaluating the current protocols and mechanisms, we can identify areas for improvement, such as reducing the number of messages exchanged, minimizing data transfer sizes, and streamlining the consensus process. These optimizations enable faster communication between nodes, improving system responsiveness and throughput.
- Resource Scheduling. The efficient allocation of system resources, such as CPU, memory, and network bandwidth, is essential for ensuring that different tasks are executed effectively. By monitoring system load and task requirements, we can dynamically adjust the allocation of resources to meet the needs of each task. This ensures that critical tasks receive the necessary resources to complete their work quickly, while less demanding tasks do not consume excessive resources. Resource scheduling helps optimize system performance and maximize the utilization of available resources.
- Caching Strategy. The utilization of caching technology is a powerful tool for improving data access speed and reducing duplicate calculations and transmissions. By storing frequently accessed data in a cache, we can quickly retrieve it without having to access the underlying data store. This significantly reduces the latency of data access operations, improving system responsiveness. Caching also reduces the need for redundant calculations and transmissions, saving system resources and improving overall performance. Implementing an effective caching strategy tailored to the specific needs of the system is crucial for achieving optimal performance.

In the realm of Byzantine fault-tolerant systems, wireless networks play a pivotal role in enabling seamless information exchange between a diverse array of applications. However, the inherent challenges of wireless communication channels, including fading, interference, and their open nature, pose significant obstacles.

Fading is a natural phenomenon that occurs in wireless communication channels, causing the signal strength to attenuate over distance. This attenuation increases the likelihood of packet loss or damage during transmission, introducing a higher degree of information inconsistency in inter-node communication within Byzantine fault-tolerant systems. As a result, consensus formation, a crucial component of these systems, faces greater challenges in achieving both efficiency and accuracy.

Interference in wireless networks can arise from various sources, such as other wireless devices or electromagnetic interference. This interference reduces the overall communication quality, potentially leading to nodes being unable to receive or send information correctly. In Byzantine fault-tolerant systems, where the detection and identification of malicious or faulty nodes is paramount, interference can significantly increase the complexity of this task, making it more difficult to ensure the integrity and reliability of the system.

The openness of wireless channels makes them inherently vulnerable to malicious attacks. Attackers can exploit this openness to launch various types of attacks, such as forging, tampering, or replaying data packets. These attacks can compromise the security of Byzantine fault-tolerant systems, potentially leading to system crashes, data leaks, or even the manipulation of consensus outcomes. Therefore, it is crucial to implement robust security measures to protect wireless networks and ensure the resilience of Byzantine fault-tolerant systems against such threats.

The choice of modulation technology determines the transmission method of signals in wireless channels, which affects the transmission efficiency and reliability of signals. Inappropriate modulation techniques may cause signal distortion or attenuation during transmission, increasing the risk of packet loss or damage, thereby increasing the possibility of information inconsistency in communication between nodes in Byzantine fault-tolerant systems, and affecting the efficiency and accuracy of consensus formation.

Encoding technology is used to add redundant information during data transmission to improve the reliability of data transmission. In the Byzantine fault-tolerant system, encoding technology can effectively detect and correct errors in data transmission, reducing the risk of packet loss or damage. However, if the encoding technology is chosen improperly or the encoding redundancy is insufficient, it may lead to a decrease in error detection and correction capabilities, increase the difficulty of detecting and identifying malicious or faulty nodes in the system, and reduce the system's fault tolerance and security.

The size of information capacity directly affects the transmission capacity and system scalability of wireless networks in Byzantine fault-tolerant systems. When the information capacity is insufficient, the system may not be able to support a large amount of data transmission, resulting in increased data transmission latency and extended consensus formation time. In addition, if the information capacity is insufficient, the system may not be able to accommodate more nodes to join, limiting the scalability of the system. Therefore, in the Byzantine fault-tolerant system, it is necessary to fully consider the demand for information capacity to ensure that the system can support sufficient data transmission and node expansion.

Designing a reasonable Byzantine fault-tolerant consensus protocol can solve the above problems [11–13]. Byzantine fault-tolerant consensus is a key aspect of Byzantine fault-tolerant systems, ensuring that network nodes can reach consensus on consistent values even in the presence of faulty or damaged components. The Byzantine fault-tolerant consensus protocol occupies a pivotal position within the Byzantine fault-tolerant system, serving as a crucial mechanism for ensuring the security and consistency of distributed systems. Specifically, its role within the Byzantine fault-tolerant system can be elaborated from several perspectives.

The Byzantine fault-tolerant consensus protocol is indeed a cornerstone of the Byzantine fault-tolerant system. In distributed systems, where numerous nodes work collaboratively to maintain the system's functionality, it is imperative that they communicate effectively and coordinate their actions. However, the presence of potential node failures or malicious attacks introduces unknown risks and uncertainties, posing a significant challenge to the system's stability. To mitigate these risks, the Byzantine consensus protocol employs sophisticated algorithms and mechanisms. Its purpose is to enable all honest nodes to converge on a common understanding or "consensus" in an environment where failures and malicious behavior are ever-present. By doing so, the protocol safeguards the system's correctness and reliability, ensuring that it can continue to operate smoothly despite the challenges it faces.

The Byzantine fault-tolerant consensus protocol tackles the crucial issue of fault tolerance in distributed systems head-on. It employs cryptographic techniques, which are robust methods for secure communication and data integrity, and the majority voting principle, a democratic approach that relies on the majority of honest nodes to make decisions. Through these techniques, the protocol achieves consensus even when unreliable and malicious nodes are present in the system. Furthermore, the protocol involves multiple rounds of negotiation and message passing, a process that allows it to identify and isolate those nodes that exhibit malicious behavior, thus preventing them from compromising the integrity of the entire system. Additionally, the protocol utilizes fault-tolerant data replication techniques, ensuring that even if some nodes fail or are attacked, other nodes can still provide accurate replicas of data, maintaining the system's data reliability and consistency.

In addition to focusing on ensuring system correctness and consistency, the Byzantine fault-tolerant consensus protocol also prioritizes performance optimization. Recognizing that high performance is crucial for practical applications, the protocol strives to achieve both correctness and speed. By optimizing its algorithms, the protocol reduces unnecessary computational overhead and ensures that nodes can efficiently process and transmit data. Furthermore, the protocol minimizes communication overhead by carefully designing its message-passing mechanisms, ensuring that only necessary information is exchanged between nodes. Finally, the protocol aims to enhance consensus speed, allowing nodes to converge on a consensus more quickly, thus improving the overall response time of the system. By balancing these considerations, the Byzantine fault-tolerant consensus protocol ensures that distributed systems can meet demanding performance requirements in real-world scenarios, effectively supporting a wide range of distributed application scenarios.

In a comprehensive summary, the Byzantine consensus protocol plays a pivotal role in the Byzantine fault-tolerant system, standing as its backbone. This protocol relies heavily on the robust foundation of network communication, cryptography technology, and precise time synchronization to guarantee the stability and security of the entire distributed system. These underlying technologies work hand in hand to ensure that the Byzantine consensus protocol can effectively detect and mitigate failures or malicious behavior among nodes. More specifically, the Byzantine consensus protocol leverages network communication to enable nodes to exchange critical information and coordinate their actions. Cryptography technology ensures that the exchanged data is secure and tamper-proof, preventing malicious nodes from altering or intercepting messages. Additionally, precise time synchronization among nodes is essential to maintain the consistency and accuracy of the consensus process. With these foundations in place, the Byzantine consensus protocol can provide robust support for various distributed applications. It ensures that all honest nodes reach a common understanding or "consensus" despite the presence of failures or malicious nodes, thus maintaining the integrity and reliability of the system. This capability is particularly crucial for secure transactions and data-sharing scenarios, where the involvement of multiple parties and the potential for malicious interference make the consensus process even more vital.

In conclusion, the Byzantine consensus protocol is an essential component of the Byzantine fault-tolerant system, leveraging the power of network communication, cryptography, and time synchronization to ensure the stability, security, and reliability of distributed systems. Its ability to provide strong support for distributed applications, secure transactions, and data-sharing scenarios further underscores its importance and value in today's interconnected world.

3.1.1 Support for Byzantine Fault-Tolerant Consensus Protocols

The support required by Byzantine fault-tolerant consensus protocol mainly includes network communication, cryptography, time synchronization, fault tolerance, robustness design, etc., which are specifically expressed and summarized as follows:

Network Communication The Byzantine consensus protocol relies heavily on robust network communication to enable nodes in a distributed system to achieve consensus on critical transactions or block writes. The foundation of this protocol is the assurance of the integrity of data transmission, the accuracy of reception, and the synchronization and coordination between nodes. In the process, nodes engage in a series of communication exchanges, including pre-preparation, preparation, and submission messages, utilizing the majority principle—typically requiring more than two thirds of nodes to agree—to achieve consensus. The reliability of network communication is paramount for the Byzantine consensus protocol, as it must withstand potential attacks from malicious nodes and guarantee that the system can maintain normal operation and achieve consensus, even in the presence of adversarial nodes. This robust communication framework ensures that messages are securely transmitted and accurately received, thereby enabling the consensus mechanism to function effectively.

Cryptography The Byzantine consensus protocol utilizes cryptography techniques to safeguard the reliability of inter-node communication and data integrity in distributed systems. By employing a secure information transmission mechanism, the protocol ensures that messages are both authentic and tamper-resistant. Specifically, digital signature technology is utilized to authenticate the source and content of messages, thus guaranteeing their authenticity and preventing any unauthorized modifications. Additionally, a hash function is employed to generate a unique hash value for each message, serving as a verification basis for data integrity. Any minor changes to the message will result in changes to the hash value, which will be detected by the system. Furthermore, encryption technology is used to encrypt the communication content, safeguarding the security of information during transmission and preventing it from being intercepted or tampered with by unauthorized parties. The comprehensive application of these cryptographic techniques enables the Byzantine consensus protocol to implement a robust consensus mechanism in distributed systems, ensuring system consistency and security, even in environments where malicious nodes are present.

Time Synchronization The Byzantine consensus protocol plays a crucial role in ensuring consistency and security between nodes in distributed systems, with time synchronization playing a crucial role. First, in the Byzantine consensus agreement, nodes need to exchange a series of messages to reach a consensus. Time synchronization ensures that nodes can receive and process these messages in the correct order, avoiding message sequence confusion caused by network latency or malicious node interference. By using timestamps or other time stamps, nodes can accurately identify the sending and receiving times of messages, thereby ensuring the sequence and consistency of messages. Secondly, malicious nodes may attempt to disrupt the normal operation of the Byzantine consensus protocol by delaying message delivery, which is known as a denial-of-service (DoS) attack. However, time synchronization enables nodes to recognize this type of attack. Because malicious nodes are unable to alter the time of other normal nodes, they are unable to effectively delay message delivery and disrupt protocol execution. Finally, under the synchronization mechanism, nodes can operate and make decisions according to a predetermined schedule. This means that nodes do not need to spend extra time waiting for other nodes, thereby improving the operational efficiency and response speed of the system. Time synchronization ensures that all nodes operate within the same time frame, enabling the system to reach consensus and respond to external requests more quickly.

Fault Tolerance and Robustness Design The Byzantine consensus protocol relies on fault-tolerant and robust design to ensure that consistent decisions can be made and the stable operation of the system can be maintained even in the presence of faulty or malicious node attacks in distributed systems. First, fault tolerance design is reflected in the system's ability to tolerate a certain number of node failures or malicious behavior without affecting the correctness of the entire system. For example, in the Byzantine fault-tolerant (BFT) protocol, as long as more than two thirds of the nodes are honest, the system can ensure consistent decisions through the majority principle. Secondly, robustness design focuses on the stability and reliability of the system in the face of various abnormal situations, such as network latency and packet loss. The Byzantine consensus protocol ensures the reliability and security of communication between nodes in complex environments by introducing fault detection mechanisms, message signature verification, and adopting multiple communication protocols. At the same time, the protocol also improves the efficiency and response speed of the system by optimizing algorithms and reducing additional overhead. In summary, the Byzantine consensus protocol ensures that distributed systems can maintain consistency and security in the face of various challenges through fault-tolerant and robust design.

3.1.2 Applications

The functions supported by Byzantine fault-tolerant consensus protocol are distributed applications, secure transactions, data sharing and blockchain, etc., which are specifically expressed and summarized as follows.

Distributed Applications The Byzantine consensus protocol effectively supports the operation of distributed applications through its fault tolerance and robustness design, as well as its unique consensus mechanism. Specifically, it ensures that in distributed networks, even in the presence of malicious nodes or network failures, each node can still reach consensus, maintaining system consistency and reliability. In distributed applications, the Byzantine consensus protocol ensures that all participating nodes maintain synchronization and consistency in data exchange, state updates, and decision making, thereby avoiding issues of data inconsistency and decision conflicts. In addition, the Byzantine consensus protocol also provides efficient messaging and verification mechanisms, as well as a flexible network topology, enabling distributed applications to better adapt to complex and ever-changing network environments, achieving efficient, stable, and secure operation.

Secure Transactions The Byzantine consensus protocol supports secure transactions by ensuring that secure and reliable consensus can be reached even in the presence of unreliable and malicious nodes in distributed networks. This agreement is based on the principle of majority voting and requires consensus to be reached among a certain number of nodes, even if there are up to (n-1)/3 malicious nodes, the system can still maintain consistency in its state. During the trading process, the proposer broadcasts the trading proposal to other nodes in the network, which validate and vote on the proposal. The transaction will only be confirmed and executed when it receives sufficient approval from the correct nodes, thereby ensuring the security and consistency of the transaction. This mechanism effectively solves the Byzantine fault-tolerant problem in distributed networks, enabling secure and reliable transactions in complex and unpredictable network environments.

Data Sharing The Byzantine consensus protocol effectively supports data sharing through its unique fault-tolerant mechanism and distributed network architecture. First, it allows nodes in the network to reach consensus on the status and updates of data without the need for centralized trust institutions. Secondly, the Byzantine consensus agreement adopts the principle of majority voting, ensuring that in the presence of faulty or malicious nodes, consensus decisions can still be reached through the correct nodes. This makes data sharing in distributed networks more secure and reliable. At the same time, the protocol can also handle complex situations such as network latency and asynchronous communication, ensuring the efficiency and real-time performance of data sharing. In summary, the Byzantine consensus protocol provides a solid foundation and effective support for data sharing with its strong fault tolerance and decentralized characteristics.

Blockchain The Byzantine consensus protocol effectively supports the operation of blockchain through its unique fault-tolerant mechanism and decentralized characteristics. Specifically, the Byzantine consensus protocol allows for a distributed approach to reach a quorum or a sufficient group of nodes to reach consensus in a network environment with potential malicious nodes or Byzantine generals. In this process, each node can independently determine the trusted object without relying on the consistent behavior of all nodes. In addition, variant protocols such as improved practical Byzantine fault tolerance and the authorized Byzantine Fault tolerance (ABFT) algorithm further enhance the fault tolerance and security of blockchain systems. PBFT ensures the performance and security of the system even in limited node situations while tolerating a certain number of Byzantine nodes, through methods such as electing master nodes, broadcasting transactions across the network, and verifying hash digest. ABFT, on the other hand, selects bookkeepers based on equity ratios and achieves consensus through Byzantine fault-tolerant algorithms, ensuring that each block has finality and prevents forking. In summary, the Byzantine consensus protocol and its variants provide a strong consensus mechanism for blockchain systems, ensuring the security, transparency, and immutability of blockchain data.

3.2 Attacks on Byzantine Fault-Tolerant Consensus Protocols

"Byzantine failures" represent a pivotal challenge in achieving consensus within distributed systems. These failures refer to the potential for nodes within a distributed network to exhibit unpredictable and malicious behavior, including the dissemination of contradictory or falsified information to other nodes. The common attacks in Byzantine fault-tolerant consensus are mainly divided into four layers: physical layer attacks, protocol layer attacks, data layer attacks, and cross-layer attacks.

3.2.1 Physical Layer Attacks

Byzantine attacks targeting the physical layer in fault-tolerant consensus systems represent a significant security threat. These attacks focus on compromising the underlying physical components of a distributed communication system, ultimately aiming to disrupt or manipulate the vital communication process between nodes. Such attacks derive from the Byzantine generals problem, a theoretical framework in distributed computing that explores the challenges of achieving consensus among nodes in the presence of faulty or malicious participants. When it comes to Byzantine attacks targeting the physical layer, several common techniques are employed by malicious actors. One such technique is physical tampering, which involves the deliberate modification or destruction of hardware components. This can range from simple acts of vandalism to sophisticated methods of hardware hacking, all aimed at compromising the integrity and functionality of the system. Another technique is signal jamming, which involves the use of radio frequency interference or other means to disrupt or block communication signals.

By drowning out or overwhelming legitimate signals, attackers can prevent nodes from communicating effectively, thereby disrupting the consensus process. Signal spoofing is another common technique used in Byzantine attacks targeting the physical layer. This involves the creation of fake or misleading signals that mimic legitimate communication. By sending spoofed signals, attackers can trick nodes into accepting incorrect information, causing them to make incorrect decisions or even become isolated from the rest of the system. The successful mitigation of Byzantine attacks targeting the physical layer requires a multilayered defense strategy that combines robust hardware design, secure communication protocols, and effective monitoring and detection mechanisms. By understanding the techniques and motivations behind these attacks, system designers and operators can better protect their distributed systems from malicious actors seeking to disrupt or manipulate the communication process.

I. Physical Tampering Physical tampering is a highly sophisticated and dangerous technique employed by malicious actors to disrupt or compromise the communication process in distributed systems. This technique involves the deliberate manipulation or modification of critical hardware components, including, but not limited, to antennas, receivers, and transmitters. In a Byzantine attack targeting the physical layer, physical tampering can take many forms. For instance, attackers might physically damage or alter antennas, weakening their ability to transmit and receive signals. This can lead to signal loss, reduced communication range, or even complete communication failure. Similarly, receivers and transmitters, the heart of any communication system, can be tampered with. Attackers might install malicious hardware that intercepts or modifies signals, causing nodes to receive incorrect or misleading information. This can confuse the consensus process, leading to incorrect decisions or even the complete breakdown of the system. Physical tampering can also involve the insertion of malicious code or firmware into hardware components. This code can be triggered remotely, allowing attackers to gain control over the communication process and potentially execute arbitrary commands on the compromised nodes. To mitigate the risk of physical tampering, system designers and operators must employ robust hardware security measures. This includes using tamper-resistant enclosures, securing sensitive components with encryption, and regularly monitoring and auditing hardware for signs of tampering. Additionally, secure supply chain management practices can help ensure that malicious hardware is not introduced into the system during manufacturing or distribution. By understanding the threat posed by physical tampering and taking appropriate security measures, distributed systems can better defend against Byzantine attacks targeting the physical layer and maintain the integrity and availability of critical communication processes.

The impacts of physical tampering on Byzantine fault-tolerant consensus are as follows.

- Direct Impact on Node Status. The direct impact of Byzantine fault-tolerant consensus attacks on node states is significant, as it may lead to nodes receiving incorrect or inconsistent information, thereby disrupting trust relationships between nodes, making node states unstable and difficult to predict. This type of attack not only increases the difficulty of verifying the authenticity and accuracy of messages by nodes but may also reduce the performance and security of the entire system, and even trigger the risk of network fragmentation, seriously threatening the stable operation of blockchain networks. For example, Zhou et al. [14] presented an enhanced BFT algorithm known as verification group (VG) RAFT, aiming to bolster Byzantine fault tolerance and consensus efficiency through mechanisms like digital signatures, in reaction to certain existing improved algorithms. However, it still exhibits limitations in certain performance metrics. This approach incorporates the notion of node trust values, allowing the node cluster to select those with higher trust scores to constitute a validation group. This validation group is tasked with scrutinizing the consensus process to prevent erroneous logs from being appended to the blockchain.
- Destruction of Communication Channels. Physical tampering may also include the destruction of communication lines or devices between nodes, resulting in communication interruption or interference between nodes. This will affect the smooth progress of message passing and consensus processes between nodes. For example, Wu et al. [15] introduced a novel small-sample collaborative spectrum sensing approach known as weighted sequence detection (WSD) to mitigate the threat posed by large-scale Byzantine attacks. This method initially leverages channel state information to authenticate the robustness of global decisions formulated by the fusion center and assess the trustworthiness of individual sensors. Furthermore, the methodology incorporates the weights of each sensing sample, which are tied to the trust values, into the sequential detection process within a defined time window. To enhance the efficiency and accuracy of the system, a sequential arrangement based on the descending order of trust values is also proposed.
- Reduce System Security. Physical tampering allows attackers to bypass system security mechanisms and directly operate on hardware or data. This will reduce the overall security of the system and may render the Byzantine fault-tolerant consensus mechanism ineffective. For example, Alexander et al. [16] considered that the distributed nature of wireless sensor networks increases their relative vulnerability to physical tampering and disruption. Therefore, the authors addressed this emerging vulnerability by introducing blockchain applications, learning the correlations between low dimensional projections of observation sequences captured by paired sensors, and using these correlations as the baseline for a soft consensus mechanism that can identify damaged sensors with the strongest paired statistical anomalies.

II. Signal Jamming Signal jamming, a technique employed to disrupt or hinder communication, involves the deliberate transmission of strong interfering signals. These jamming signals are specifically designed to overwhelm or block legitimate communication signals, creating interference that makes it difficult or impossible for receivers to correctly decode and process the intended information. Such jamming can have serious consequences, ranging from minor inconveniences like disrupted phone calls to more significant issues like the disruption of critical communication systems in emergency situations.

The impact of signal jamming on Byzantine fault-tolerant consensus are as follows:

- Reduced Communication Reliability. Byzantine fault-tolerant consensus relies on reliable communication between nodes to achieve consensus. Signal interference may cause delay, loss, or damage in message transmission, making it difficult for nodes to receive information from other nodes in a timely and accurate manner. This will directly affect the efficiency and accuracy of the consensus process. For example, in real-world applications, wireless blockchain systems face inherent challenges due to constrained physical resources and unreliable wireless channels caused by environmental noise and potential adversarial attacks. To address these issues, Zhang et al. [17] explored the design of a Byzantine fault-tolerant consensus protocol for blockchain operations in single-hop wireless networks, taking into account signal-to-noise ratio constraints. Their proposed solution, named LRBP, is a low-latency and reliable Byzantine fault-tolerant consensus protocol that has been proven to satisfy both persistence and activity properties, ensuring the robustness and efficiency of the blockchain network.
- Increased Difficulty of Consensus. In Byzantine fault-tolerant consensus, it is usually necessary for more than half of the nodes to reach a consensus to ensure the correct operation of the system. However, signal interference may result in some nodes being unable to receive or recognize information from other nodes, thereby increasing the difficulty of reaching a consensus. In extreme cases, it may even lead to the inability to complete the consensus process. Soesanto et al. [18] introduced an adaptive Proof-of-Work (PoW) framework design utilizing multiple member pools, which leads to variations in transaction speeds. Their approach facilitates the customization of consensus difficulty levels for each transaction. This adjustment in PoW difficulty is achieved by analyzing and processing transaction data to determine the significance or importance of each transaction.
- System Performance Degradation. Signal interference not only affects the efficiency and accuracy of the consensus but may also have a negative impact on the performance of the entire distributed system. For example, nodes may need to spend more time waiting for message transmission and confirmation, resulting in an increase in system response time. Meanwhile, frequent communication errors may also lead to wastage of system resources and performance degradation. For example, Moniz et al. in [19] delved into the integration of Byzantine nodes and

dynamic transmission faults that can potentially impact the communication flow between non-Byzantine nodes. To address this, they propose a model that builds upon the communication failure model presented by Santoro and Widmayer. Their model acknowledges the presence of dynamic and transient transmission faults, assuming that communication between nodes can be faulty at one instant and reliable at another. In a wireless context, this signifies that a broadcast message may not be uniformly received by its intended recipients. Some may successfully receive the message, while others may fail to do so. Under extreme conditions such as a jamming attack, all messages may even be lost for a period of time

- Increased Security Risks. In systems with malicious nodes, signal interference may provide opportunities for malicious nodes to take advantage of. Malicious nodes may use signal interference to interfere with the communication of normal nodes, making them unable to participate in the consensus process or receive correct information. This will further increase the security risk of the system and may lead to serious issues such as data inconsistency or system crashes. Lei et al. [20] introduced the reputation-based Byzantine fault-tolerant algorithm, incorporating a reputation model to assess the performance of each node during the consensus process. If malicious behavior is detected, the offending node's voting power and reputation are diminished. Furthermore, they presented a novel reputation-driven primary change scheme, where nodes with higher reputations have a greater likelihood of becoming master nodes and generating new, valid blocks. This approach mitigates the security risks associated with the master node.

III. Signal Spoofing Signal spoofing, as a form of malicious activity in distributed systems, refers to the deliberate action of certain nodes to disrupt normal communication and consensus processes by sending incorrect or false information to other nodes. These malicious nodes aim to mislead and confuse the system by transmitting forged, tampered, or delayed messages. Such attacks can have significant consequences, ranging from the degradation of system performance to the complete disruption of critical services. Signal spoofing attacks can be difficult to detect and mitigate, requiring robust security measures and vigilant monitoring of the system's communication channels.

The impact of signal spoofing on Byzantine fault-tolerant consensus are as follows:

- Breaking Consensus. Signal spoofing may result in loyal nodes receiving incorrect information, making it impossible to reach a consensus. If malicious nodes can successfully deceive enough loyal nodes, they may dominate the consensus process, leading to instability or failure of the entire system. Melih et al. [21] introduced a cutting-edge blockchain-driven approach to thwart incoming call display spoofing attacks in a real-time environment. Their solution employs a low-latency consensus algorithm to oversee and authenticate caller ID data from Internet service providers (ISPs) and institutions. A two-step verification process is outlined, ensuring the accuracy and integrity of Automatic Number Identification (ANI) information at crucial stages of the call. Additionally, their method initiates an update of ISP registrations whenever the caller ID changes, thus rendering it impervious to issues such as roaming, use of Internet Protocol Private Branch Exchange (IP-PBX), Virtual Private Network (VPN), and other abnormal scenarios.

- Reduced System Availability. Due to the interference of malicious nodes, loyal nodes may need to spend more time and resources verifying and filtering false information, thereby reducing the overall availability of the system. Nagananda et al. [22] considered that one out of N sensors was subjected to data tampering attacks from Byzantine sensors, where intruders injected higher noise power into the measurements of the attacked sensors (compared to unconnected sensors). Their main contribution is to analyze the impact of the aforementioned types of data tampering attacks on the performance of the rank-reduced LS estimator. Especially, they demonstrated how to optimize the priority of information-rich content in data in the presence of attacks.
- Increased Security Risk. Signal spoofing may make the system more susceptible to other types of attacks, such as dual payment attacks and denial-of-service attacks. Malicious nodes may use false information to conceal their true intentions, thereby increasing the security risk of the system. For example, Santoro et al. [23] studied transient and common faults in synchronous distributed systems. This includes systems where any communication link fails, systems where each processor sends or receives a fault once or another, and systems where normal functionality is restored after a limited time after a fault. Please note that traditional component failure models cannot handle these situations. The model they use is the communication fault model, also known as the transmission fault or dynamic fault or mobile fault model. Using this model, the basic consistency problem of any topology synchronous network with common faults was studied.

3.2.2 Protocol Layer Attacks

At the protocol layer, Byzantine attacks represent a significant threat to the fundamental principles of network communications, namely integrity, availability, and confidentiality. These attacks capitalize on vulnerabilities within the communication protocols, exploiting weaknesses to undermine the reliability and security of the entire network. Some of the most common examples of Byzantine attacks at this level are sybil attacks, which involve the creation of numerous fake identities to flood the network with illegitimate messages. Injection attacks involve the malicious insertion of false or corrupted data into the communication stream, disrupting the accuracy of information exchange. Forking attacks aim to create multiple conflicting versions of the network's state, leading to confusion and inconsistencies. Finally, denial-of-service (DoS) attacks, a subset of Byzantine attacks, target network resources by overwhelming them with requests or traffic, rendering the network inaccessible to legitimate users. DoS attacks, in particular, have the potential to wreak havoc on the normal flow of communication within a fault-tolerant consensus system. When network resources are overwhelmed, legitimate users are unable to access the system, leading to significant financial and operational consequences. Such attacks can be particularly devastating in environments where consensus is critical, such as blockchain networks or distributed ledgers, where the integrity and availability of the system are paramount. To mitigate the risks posed by Byzantine attacks, robust security measures must be implemented at the protocol layer. This includes the use of encryption, authentication mechanisms, and anomaly detection systems to identify and isolate malicious nodes. Additionally, fault-tolerant consensus algorithms are designed to withstand such attacks, ensuring that the network can continue to operate effectively even in the presence of malicious actors.

I. Sybil Attacks As shown in Fig. 3.2, the sybil attack, often referred to as the witch attack, is a sophisticated network attack strategy that poses a significant threat to the security and stability of various systems. In this type of attack, an attacker employs deceptive techniques to forge multiple identities or nodes within the network, aiming to obtain illegitimate benefits or disrupt the normal operation of the system. The fundamental principle of the sybil attack lies in the attacker's ability to manipulate the system's judgment by controlling a numerically superior number of false identities or nodes. To execute a sybil attack, attackers can utilize various techniques to forge identities. One common method is by controlling multiple IP addresses, allowing them to create numerous fake nodes within the network.

Additionally, attackers can employ virtual machines or proxy servers to disguise their true identities and appear as legitimate users. These false identities can then be used to participate in various activities within the system, including voting, messaging, and other consensus mechanisms. The types of sybil attacks can be broadly classified into direct and indirect communication attacks. Direct

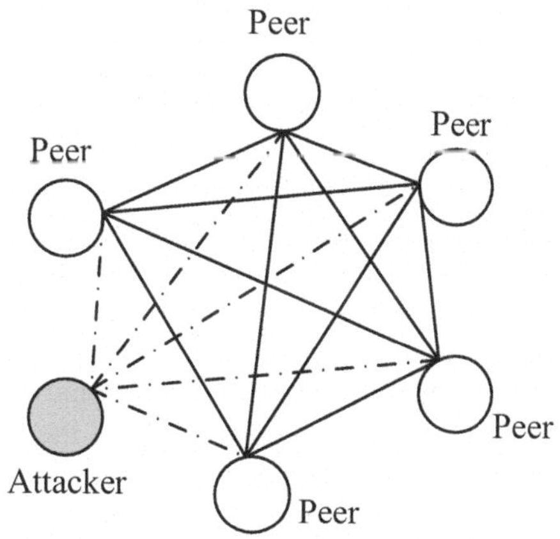

Fig. 3.2 The principle of sybil attack based on blockchain network

communication attacks involve the attacker directly interacting with other nodes using their forged identities. Indirect communication attacks, on the other hand, involve the attacker manipulating the communication between legitimate nodes, often by intercepting and modifying messages. Furthermore, sybil attacks can be categorized based on the nature of identity forgery. Identity forgery attacks involve the creation of entirely new false identities, while identity theft attacks involve the impersonation of existing legitimate users. Simultaneous attacks refer to attacks where multiple forged identities act simultaneously to achieve a specific objective, while nonsimultaneous attacks involve staggered or staggered actions. It is crucial to recognize and mitigate the risks posed by sybil attacks to ensure the security and reliability of network systems. Implementing robust authentication mechanisms, monitoring network traffic, and deploying anomaly detection systems can help identify and isolate attackers, thereby minimizing the impact of these devastating attacks.

Direct Communication Sybil nodes communicate directly with legitimate nodes. For example, Jiang et al. [24] considered that PoW-based systems generate high latency and low throughput, making them ineffective in handling real-time Internet of Things (IoT) applications. On the other hand, due to susceptibility to sybil attacks, the Byzantine fault-tolerant (BFT) consensus algorithm with better latency and throughput performance cannot be used in permissionless settings. They present a sybil-proof wireless network coordinate-based Byzantine consensus (SENATE), which has the merits of both real-time consensus reaching and sybil-proof, i.e., it is based on the conventional BFT consensus framework yet works in open systems of wireless devices where faulty nodes may launch sybil attacks.

Indirect Communication There are no legitimate nodes that can directly communicate with sybil nodes, but instead route and forward through one or more malicious nodes. Thawani et al. [25] delved into the security aspects of the Temporary Ordered Routing Algorithm (TORA) in the context of sybil attacks. TORA, which relies on a series of link reversal algorithms, is a highly adaptive distributed routing mechanism employed in mobile ad hoc networks (MANETs). It offers multiple loop-free routes to any destination through its route creation, maintenance, and erasure functionalities.

Forgery of Identity and Identity Theft When an attacker generates an arbitrary sybil identity, while identity theft is when the attacker assigns a legitimate identity to a sybil node. Michail et al. [26] introduced an active strategy leveraging social network data, primarily aimed at thwarting identity spoofing in online sub-communities embedded within larger social ecosystems, like Facebook groups or subsets. This approach is adaptable to diverse social media applications and boasts a high degree of accuracy in detecting deceptive accounts seeking to infiltrate these sub-communities; Tan et al. [27] delved into a novel denial-of-service attack in dynamic spectrum access networks, dubbed the sybil-enabled attack. In this sophisticated approach, the assailant not only executes a primary user impersonation (PUE) attack but also devises and infiltrates numerous sybil identities through Byzantine-style assaults, thereby aiming to disrupt the decision-making mechanisms of the secondary network. To substantiate the feasibility and potency of this attack, it was executed on a cognitive radio test bench. Additionally, the researchers analyzed the optimal attack strategy from the malevolent attacker's perspective, focusing on the ideal allocation of sybil interfaces for varying attacks, to maximize the adverse impact on the secondary network.

Simultaneous Attack Refers to the attacker's simultaneous participation of all their sybil identities in a single network communication, rather than a simultaneous attack where the attacker only uses a portion of their sybil identities within a specific time period. Attackers may adopt the guise of multiple vehicles traveling concurrently to execute Sibyl attacks. Such attacks become particularly menacing when vehicles collude with others to authenticate sybil vehicles using valid credentials. To counter this, Rabieh et al. [28] introduced a cross-layer solution that allows roadside units (RSUs) to identify these sybil vehicles. Their approach hinges on verifying the vehicle's location, as a sybil vehicle would not be in its claimed position. To detect a vehicle's presence, directional antennas are employed to transmit challenge packets to the vehicle's stated location. If the vehicle is indeed in the anticipated position, it should be able to receive the challenge and respond with a valid packet. Cost-effectively, challenge packets are not sent continuously to all vehicles but only when suspicion of a Sibyl attack arises.

II. Injection Attacks Injection attacks refer to the deliberate insertion of malicious data into an application's input field by attackers, enabling them to gain unauthorized access or perform malicious operations. This attack method poses a significant threat to the security and integrity of a system, potentially leading to dire consequences such as data leakage, data tampering, and denial of service. When data injection attacks occur within the context of Byzantine fault-tolerant consensus, the situation becomes even more complex and dangerous. Byzantine fault tolerance is a framework that allows a system to continue functioning correctly even when some nodes in the network are faulty or malicious. However, when attackers exploit this framework through data injection attacks, they can compromise the entire system's security. In a Byzantine fault-tolerant consensus environment, malicious nodes can affect the normal operation of the system by forging, tampering with, or injecting false data. These attacks can disrupt the consensus process, leading to incorrect decisions being made by the system. For instance, attackers can inject false transaction data into the blockchain, affecting the accuracy of the ledger and potentially leading to double-spending or other fraudulent activities. To mitigate the risks posed by data injection attacks in Byzantine fault-tolerant consensus systems, it is crucial to implement robust security measures. This includes employing encryption techniques to ensure the integrity and authenticity of data, implementing access controls to restrict unauthorized access, and monitoring network traffic to detect and isolate malicious nodes. Additionally, developing and deploying anomaly detection systems can help identify suspicious patterns and activities, further enhancing the security of the system.

For example, Zaka et al. [29] introduced a technique utilizing artificial neural networks for detecting false data injection attacks in home area networks. Prompt and precise identification of these falsified measurements is vital in taking necessary countermeasures to mitigate potential threats. Dai et al. [30] considered that false data injection attacks (FDIA) disrupt information exchange between controller units and pose a significant threat to microgrids. Given the imbalance of advantages between attackers and defenders, there is an urgent need for an effective solution to enhance network resilience. To alleviate this issue, they proposed a proposed framework to enhance network resilience by leveraging the inherent security of blockchain technology, replacing fragile information exchange and computation with secure transactions. Unlike current methods in the control field with limited network resilience, this framework considers both communication and control domains. The smart contract (SC) deployed on the enterprise-level HyperLedger blockchain provides distributed secondary control and self-healing capabilities, protecting microgrid secondary control from the impact of FDIA in a zero-trust environment. Samuel et al. [31] considered the necessity of uninterrupted communication between physical and virtual twins, emphasizing that secure data sharing is paramount in human digital twin (HDT) systems. Given the significance of data security and privacy, existing blockchain-based solutions often struggle to meet the stringent latency and throughput demands of HDT systems, primarily due to the intricate consensus procedures employed in traditional blockchain frameworks. To address these limitations, the researchers introduced an innovative Byzantine fault-tolerant approach that leverages reputation-enhancing fragments, tailored specifically for zero-trust HDT environments. The authors propose a practical Byzantine fault-tolerant consensus framework centered on reputation enhancement through parallel verification. This framework aims to enhance throughput and minimize latency during data sharing, while maintaining the core principles of blockchain technology, including security and decentralization.

III. Forking Attacks As shown in Fig. 3.3, forking attacks are a deliberate and malicious form of disruption that can arise in distributed systems, particularly within the realm of blockchain networks. In the context of blockchain technology, the term "forking" refers to a temporary splitting of the blockchain's ledger, resulting from nodes within the network receiving and validating different blocks during

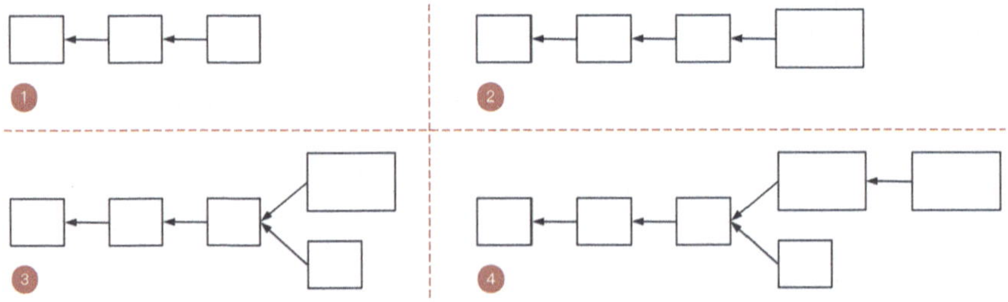

Fig. 3.3 The forking attacks

a given time frame. This splitting can occur in two main forms: hard forks and soft forks. However, when discussing forking attacks, we are primarily referring to temporary and unexpected splits that are intentionally caused by malicious actors. These attacks are typically initiated by a coordinated group of malicious nodes or miners who utilize various tactics to disrupt the network's normal operation. One such tactic is to create network congestion by flooding the system with a high volume of transactions or blocks, making it difficult for nodes to communicate and validate new blocks efficiently. This congestion can lead to delays in message delivery, causing nodes to receive and validate different blocks simultaneously. Another approach is for the malicious nodes to introduce faulty or incorrect blocks into the network, further exacerbating the splitting. These branches of the blockchain may contain different transaction sets and blocks, resulting in multiple versions of the ledger existing simultaneously. This fragmentation disrupts the Byzantine consensus mechanism, which relies on nodes agreeing on a single, unified ledger. The consequence of these forking attacks is a significant threat to the consistency and fault tolerance of the blockchain network. The splitting of the ledger creates confusion and uncertainty among participants, potentially leading to double-spending, fraud, and other malicious activities. Therefore, it is crucial for blockchain networks to implement robust security measures to detect, prevent, and mitigate the risks posed by forking attacks.

The goal of forking attachments include:

- Double pending attack: Attackers attempt to double-spend the same digital currency on a forked blockchain, confirming a transaction simultaneously on two different branches. Rezky et al. [32] examined the dangers posed by the utilization of USB flash drive devices as attack vectors. In their study, they employed a fork bomb, implemented on an Arduino Pro Micro device disguised as a USB flash drive, as the threat source. The fork bomb's objective is to degrade the memory performance of the targeted device, ultimately reducing the efficiency of executing processes. This method of utilizing USB flash drives to deliver the fork bomb attack can lead to users being unable to access the compromised operating system.
- Reducing network security: By creating multiple branches, attackers may attempt to compromise the security of the entire network, making it more vulnerable to other types of attacks. Wang et al. [33] presented a novel mechanism for confirming the most trusted chain to safeguard against fork attacks. Initially, a theoretical analysis was conducted to understand the success probability distribution of fork attacks, and a model was proposed to estimate the expected benefits of such attacks. Subsequently, an arbitration framework was introduced, encouraging branches to compete among themselves. This arbitration system delays the confirmation of the most trusted chain until the competition concludes. A termination threshold was established to mark the end of the competition, ultimately allowing the most trusted chain to emerge as the victor.
- Disrupting network stability: Multiple forks can lead to network chaos, reducing transaction confirmation speed and network availability. Marcus et al. [34]

introduced Lightweight Collective Memory (LCM), a streamlined protocol designed to establish shared memory among remote application clients for detecting integrity and consistency conflicts. LCM effectively identifies rollback attacks targeting remote applications, enhances the consistency principles of fork linearization, and alerts clients of operational stability. This protocol leverages a trusted execution environment, complements it with straightforward client operations, and requires minimal constant storage on the client side.

IV. DoS Attacks DoS attack is a network attack technique aimed at preventing computers or networks from providing normal services, resulting in unavailable or severely restricted services. This type of attack aims to disrupt the normal operation of network services by consuming the resources of the target system, such as bandwidth, processing power, or storage space. The principle of DoS attacks is mainly based on two aspects: resource depletion and service unavailability. Attackers exploit system vulnerabilities or specific network tools to consume the target system's resources by sending a large number of requests or exploiting malicious code, making it unable to respond properly to requests from other legitimate users. Attackers typically send a large number of invalid or carefully crafted requests to the target system, which may lead to system resource depletion or trigger abnormal behavior.

There are various types of DoS attacks, including, but not limited, to the following:

- Ping Flood. Attackers send a large number of Ping packets to the system, causing system resources to be heavily occupied and unable to respond to other requests. For example, Bogdan et al. [35] delved into the implications of consistent denial-of-service (DoS) attacks targeting Byzantine fault-tolerant protocols. While theoretically, the BFT protocol is designed to withstand up to one third of adversarial nodes in partially synchronized environments, its practical deployment hinges on optimistic assumptions such as instantaneous message processing and limitless memory for maintaining message queues. However, these assumptions are exposed as weaknesses when malicious nodes unleash floods of consensus messages. To address this, the researchers introduced a novel protocol for Byzantine node DoS (BDoS) attacks and successfully proved its viability within prevalent blockchain frameworks.
- SYN Flood. As shown in Fig. 3.4, attackers send a large number of SYN request packets without completing the three-way handshake process, which occupies server resources and prevents them from processing normal connection requests. Zebari et al. [36] examined and assessed the effects of HTTP and SYN flow attacks on contemporary and broadly adopted Web servers. Their performance analysis encompassed Internet Information Service 10.0 (IIS 10.0) running on Windows Server 2016 and Apache 2 deployed on a Linux Ubuntu 16.04 Long Term Support (LTS) server. The study focused on key performance metrics such as average response time, average CPU utilization, and standard deviation, which reflect the responsiveness, efficiency, and stability of network servers. The

Fig. 3.4 The SYN flood attacks

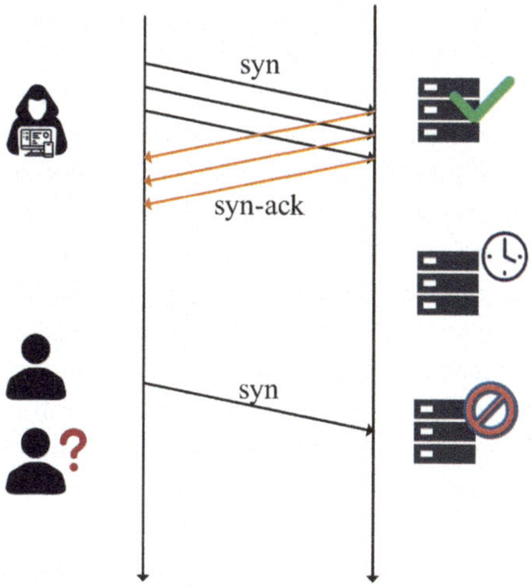

findings revealed that IIS 10.0 demonstrated higher efficiency and responsiveness in the face of HTTP flood attacks compared to Apache 2 Web servers. However, in the scenario of SYN flood attacks, Apache 2 Web servers exhibited greater responsiveness and stability.

- UDP Flood. Attackers send a large number of UDP packets to the target system, which may be invalid or forged, causing system resources to be heavily occupied. Biagioni et al. [37] devised the AllNet protocol to facilitate data transmission via UDP and other media. Upon receiving an appropriate UDP packet, an AllNet peer logs the sender's IP address and initiates the forwarding of AllNet data to that specific address. This represents a legitimate traffic amplification technique, wherein a single packet triggers the dispatch of a limited quantity (currently 100) of additional packets to the designated IP address. To mitigate the risk of attackers leveraging AllNet peers for flood amplification attacks, AllNet peers impose a verification step, requiring potential recipients to respond with a bit string sent to their respective IP addresses.

3.2.3 Data Layer Attacks

In the intricate realm of fault-tolerant consensus mechanisms, Byzantine attacks targeting the data layer of the information hierarchy pose a significant threat to the integrity and security of distributed systems. These sophisticated assaults, executed by malevolent entities, focus on manipulating or corrupting the core data and information that underpins the distributed network. The ultimate aim of these attacks is to undermine the consensus process, compromise the overall system security, or mislead other participating nodes. At this critical level, Byzantine attacks can manifest in numerous ways, each posing a unique challenge to the resilience of the distributed system. One such manifestation is data tampering, where attackers create numerous fake identities or nodes within the network. By doing so, they gain undue influence over the consensus process, skewing the collective decision-making toward their own malicious ends. Another common tactic is false reporting, where malicious nodes deliberately provide incorrect or misleading information to the consensus mechanism. This can disrupt the normal functioning of the consensus, causing delays, inconsistencies, or even complete failures. Additionally, it allows malicious actors to gain unfair advantages by misleading other participants or influencing the system's decisions in their favor. Furthermore, data injection is a particularly dangerous form of Byzantine attack. Here, attackers inject malicious data or falsified information directly into the system, often without detection. This injected data can deceive other nodes, skewing their views of the network's true state and influencing the consensus outcome in ways that favor the attackers. Such attacks can have far-reaching consequences, including the compromise of sensitive information, the disruption of critical services, and the degradation of system reliability and performance.

I. Data Tampering Data tampering is a deliberate and malicious activity that entails unauthorized or illicit alteration, deletion, or fabrication of data in order to distort its original state or purpose. This form of malicious manipulation can have profound consequences, particularly in the realm of Byzantine fault-tolerant consensus, where data integrity is crucial for maintaining the security and reliability of distributed systems. In a Byzantine fault-tolerant consensus environment, data tampering poses a significant threat. Such unauthorized modifications can lead to the loss of trust among participants, as the altered data can no longer be relied upon to accurately represent the state of the system. This lack of trust can have a cascading effect, compromising the reliability of the entire network and its ability to reach consensus on critical decisions. Moreover, data tampering can have a devastating impact on organizations and individuals who rely on the accuracy and integrity of the data within the system. It can lead to incorrect business decisions, misallocation of resources, and even legal liabilities if the tampered data is used to make decisions that have legal implications. To mitigate the risk of data tampering, organizations employing Byzantine fault-tolerant consensus mechanisms must implement robust security measures. This includes encrypting data, authenticating participants, and employing techniques such as cryptographic signatures and hashes to ensure the integrity of data. Additionally, regular audits and monitoring of the system are essential to detect and respond to any attempted data tampering in a timely manner. In conclusion, data tampering is a grave security concern in Byzantine fault-tolerant consensus environments, and it is crucial to take the necessary measures to protect the integrity and reliability of the data within these systems.

There are various types of data tampering, including, but not limited, to the following:

- Message tampering: During the message transmission process, Byzantine nodes may tamper with the content of messages sent by other nodes, causing them to lose their original meaning or become inaccurate. Take the previous work [38] as an example, Chen et. al. focus on the data tampering attack that occurs at the data layer. This attack can inflict significant harm on the crucial signature and verification processes within a blockchain network. Overall, Byzantine attacks targeting the data or information layer represent a grave threat to the integrity and reliability of distributed systems. Consequently, implementing preventive measures and designing robust consensus mechanisms are paramount in mitigating these attacks and safeguarding the accuracy of shared information across the network.

- Double voting: Byzantine nodes may engage in dual voting within the same consensus process, sending different voting results to different nodes in an attempt to disrupt consensus. Lin et. al. [39] conducted a study on secure data sharing mechanisms, encompassing mixed encryption techniques and blockchain-based secure storage solutions. Furthermore, they explored strategies for identifying and recovering tampered experimental data, leveraging an enhanced practical Byzantine fault-tolerant consensus algorithm. This research addresses crucial security concerns like data leakage and unauthorized modification in the context of experimental data sharing. The study proposes a comprehensive framework for the secure storage, sharing, and tamper-proof recovery of test data, aiming to resolve the inherent tension between safeguarding sensitive data and facilitating its sharing. This framework serves as a foundation for enabling the secure application of blockchain technology in experimental data management.

II. False Reporting False reports, a malicious and intentional act, refer to the dissemination of inaccurate or entirely fabricated information by rogue nodes within the Byzantine fault-tolerant consensus process. This behavior is designed to undermine the fundamental consistency and stability of the distributed system, ultimately jeopardizing its ability to function effectively. The aim of releasing false reports is to disrupt the consensus mechanism, which relies on honest nodes agreeing on the state of the system. When false reports are introduced, they inject doubt and confusion into the consensus process, making it difficult for other nodes to ascertain which information is genuine and trustworthy. This can have serious consequences, ranging from delays in reaching consensus to the adoption of incorrect decisions. In extreme cases, if there are a sufficient number of false reports or if they are crafted with sufficient sophistication, they may effectively prevent the system from reaching consensus altogether. This can lead to system crashes, as the lack of agreement among nodes causes the network to become unstable and potentially unusable. Alternatively, the system may experience forks, where different parts of the network adopt conflicting versions of the truth, further eroding its integrity and reliability. To mitigate the risk of false reports, Byzantine fault-tolerant consensus protocols often incorporate mechanisms such as cryptographic verification, reputation systems, and voting thresholds. These techniques help identify and isolate malicious nodes while ensuring that honest nodes have a greater influence on the consensus process. However, the threat of false reports remains a persistent challenge in distributed systems, requiring ongoing vigilance and innovation to safeguard the integrity of the consensus mechanism.

There are various types of false reporting, including, but not limited, to the following:

- Fake message: Byzantine nodes may send completely forged messages that do not match the true situation and are intended to mislead other nodes and disrupt consensus. For example, as commercial port throughput and the handling capacity of major container terminals continue to surge, the prevalence of inaccurate and misleading reporting of dangerous goods has emerged as a significant concern. Consequently, the implementation of intelligent oversight over such goods has become paramount. In response to the challenges posed by existing Practical Byzantine Fault Tolerance (PBFT) algorithms in consortium blockchain environments, such as limited scalability, high communication overhead, and inefficient performance, Lin et al. [40] have proposed a novel hybrid consensus algorithm. This algorithm, dubbed link recognition-based RAFT and PBFT hybrid consensus mechanism (LRPBFT), aims to address the aforementioned issues.
- Incorrect signature or identity forgery: In systems that require signature or identity verification, Byzantine nodes may use incorrect signatures or forged identities to send messages to deceive other nodes. Wang et al. [41] observed that prior algorithms lacked a defined range for period growth, enabling malicious nodes to manipulate their period values to gain leadership status within a RAFT cluster. This manipulation posed a threat to the fairness of leader elections. To mitigate this vulnerability, they conducted a revised attack experiment, where malicious nodes attempted to seize leadership by inflating their item values. By dynamically regulating the terminology growth range, they were able to identify and defend against these malicious nodes, thereby enhancing the effectiveness of their defense mechanism.

III. Malicious Injection In the context of Byzantine fault-tolerant consensus, malicious nodes represent a significant threat to the stability and reliability of the distributed system. These nodes, by definition, fail to adhere to the established protocol requirements and actively seek to undermine the consensus mechanism or data consistency. Malicious nodes employ various tactics to achieve their objectives. One common approach is to send incorrect or forged messages to other nodes within the network. These messages are designed to mislead honest nodes, causing them to make decisions based on false or incomplete information. The end goal is to disrupt the normal consensus process, either by preventing the system from reaching an agreement or by introducing inconsistencies to the data. The injection of malicious information can have devastating consequences for the integrity and consistency of data within the system. Data tampering or forgery by malicious nodes can compromise the trustworthiness of the information, rendering it unreliable for decision-making purposes. When other nodes rely on this corrupted data, they may make incorrect decisions that propagate throughout the network, further eroding its stability. In a Byzantine fault-tolerant consensus environment, the maintenance of data integrity and consistency is paramount. These principles form the foundation upon which the system operates, ensuring that all nodes have a shared understanding of the current state. However, the presence of malicious nodes threatens this foundation, as they can introduce errors and inconsistencies that compromise the trustworthiness of the entire system. Therefore, it is crucial to implement robust mechanisms to detect and mitigate the impact of malicious nodes. These mechanisms may include cryptographic verification, reputation systems, and voting thresholds, which help to identify and isolate rogue nodes while ensuring that honest nodes have a greater influence on the consensus process. By doing so, we can minimize the risk posed by malicious nodes and maintain the integrity and consistency of the Byzantine fault-tolerant consensus system.

There are various types of inject malicious, including, but not limited, to the following:

- SQL injection: As shown in Fig. 3.5, attackers construct malicious SQL statements and insert them into user input or other data transmission channels, causing the database to mistakenly use these malicious statements as legitimate instructions when executing SQL queries. With the proliferation of Internet users, the demand for network services and mobile/desktop network applications is surging. Consequently, the vulnerability of systems to hacking attempts is also on the rise. All Web applications rely on a backend database to store and retrieve data, which must be accessible to all users globally. Currently, SQL injection attacks pose a significant threat to Web application security, enabling attackers to steal sensitive information. Rajashree et al. [42] introduced a technique for detecting SQL injection attacks by purging parameter values from SQL queries and presented the findings of their detection method.
- Command Injection: The target of command injection attacks is command execution at the operating system level. Attackers insert malicious commands into the expected commands of the system, causing the operating system to mistakenly recognize the malicious commands as legitimate instructions when executing them. Nikhil et al. [43] delved into the challenge of addressing distributed (consensus driven) optimization problems in networks plagued by rules and malicious nodes (proxies). While standard nodes utilize distributed iterative

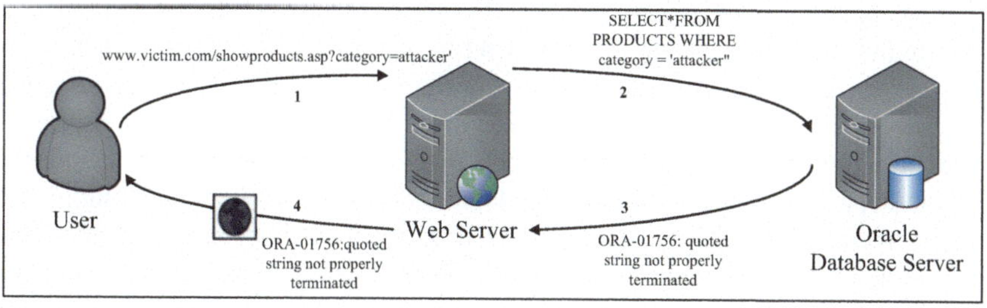

Fig. 3.5 The SQL injection

algorithms to tackle their respective optimization issues, malicious nodes attempt to steer iterations toward outcomes that align with their vested interests by injecting erroneous data. This scenario demands the detection and isolation of these harmful agents, enabling regular nodes to solve their optimization problems unhindered. The authors propose a novel approach that counters data injection attacks in distributed optimization algorithms, drawing upon the observation that malicious nodes, whether acting individually or in collusion, tend to reveal their presence through their messaging, aiming to deviate consensus values from the optimal solutions sought by conventional nodes within the network.

3.2.4 Cross-Layer Attacks

The complexity of Byzantine faults in wireless networks extends beyond just the physical, protocol, and data layers. In Byzantine fault-tolerant consensus, the convergence of attacks across multiple layers represents a significant challenge for the security and resilience of distributed systems. This cross-layer fusion of attacks manifests in the intricate interplay and coordinated efforts of Byzantine errors at various levels. The complexity of these combined attacks arises from the interconnectedness and interdependencies between different layers. For instance, physical layer attacks, such as jamming or interference, can disrupt communication channels, making it easier for protocol layer attacks to succeed. By disrupting the underlying communication infrastructure, attackers can create opportunities for protocol layer exploits, such as denial-of-service (DoS) attacks, that further compromise the consensus process. Furthermore, the fusion of attacks can also occur between the protocol and data layers.

For example, DoS attacks at the protocol layer, which aim to overwhelm the system with traffic or disrupt normal message flow, can cooperate with Byzantine node voting at the data layer. In this scenario, malicious nodes may vote incorrectly or forge messages to influence the consensus outcome, while the DoS attacks at the protocol layer provide cover or create confusion, making it more difficult for the system to detect and mitigate Byzantine faults. Similarly, attacks on the data layer, such as data tampering or forgery, can undermine the accuracy and trustworthiness of the information on which the protocol layer mechanisms rely. When malicious nodes inject incorrect or forged data into the system, it can compromise the consensus process and lead to inconsistent or incorrect decisions. This, in turn, can have cascading effects on the overall stability and performance of the distributed system. To address these cross-layer fusion attacks, it is crucial to adopt a multilayered defense approach that considers the interconnectedness and interdependencies between different layers. This includes implementing robust security mechanisms at each layer, as well as mechanisms that can detect and mitigate attacks across multiple layers. By taking a comprehensive approach, we can enhance the resilience of distributed systems against Byzantine faults and ensure their continued operation in the face of sophisticated attacks.

To effectively mitigate Byzantine attacks that traverse multiple layers, a comprehensive approach is necessary, encompassing layer-specific defenses as well as cross-layer monitoring. Li et al. [2] considered that the practical Byzantine fault-tolerant consensus mechanism has poor node scalability and is usually only used in small networks. In order to implement PBFT in large-scale IoT ecosystems and blockchain systems, they proposed a scalable consensus mechanism based on multilayer PBFT, which hierarchically groups nodes into different layers and restricts communication within the group. An optimal double-layer PBFT was proposed, which analyzed the safety threshold based on the failure probability determination (FPD) and failure number determination (FND) models and showed a significant reduction in communication complexity. In [44], Leng et al. tackled the prevalent challenges such as elevated communication overhead, compromised security, limited scalability, and tracking complexities. The author introduces a Byzantine fault-tolerant consensus algorithm built upon the double administrator short group signature framework. At the core of this algorithm, the certification authority selects the primary node and group administrator based on their respective creditworthiness. Moreover, the proposed algorithm incorporates a trace phase, enabling group administrators to trace, identify, and subsequently revoke malicious nodes based on the authentication information of both members and clients within the network. This is associated with the practical implementation of Byzantine fault-tolerant consensus across diverse layers.

Based on the above introduction, we can further summarize that most of the existing attacks in Byzantine fault-tolerant consensus can indeed be classified as physical layer attacks, protocol layer attacks, data layer attacks, and cross-layer attacks. These four types of attacks pose a serious threat to the integrity, reliability, and performance of distributed systems. The next chapter provides a detailed introduction to existing defense methods and achievements against Byzantine fault-tolerant consensus attacks.

3.3 Mitigating Attacks in Byzantine Fault-Tolerant Consensus

When it comes to wireless networks, achieving consensus among nodes is paramount for maintaining reliable and secure communication, especially when faced with the threat of malicious or faulty actors. Byzantine fault-tolerant (BFT) mechanisms have emerged as a robust solution to tackle these challenges, ensuring that the network can continue to function effectively even in the presence of such adversarial conditions. In this chapter, we delve deeper into the various defense methods that have been developed in the field of Byzantine fault-tolerant consensus. Understanding the subtle differences and applications of these methods is crucial for designing and deploying resilient wireless networks. The defense methods are primarily classified into two categories: single-layer defense and multilayer defense.

First, we focus on the Byzantine fault-tolerant consensus defense methods in single-layer wireless networks. Here, we explore the fundamental concepts, protocols, and performance evaluation techniques that underlie these defense strategies. Understanding the basic principles of these methods is essential for appreciating their effectiveness and limitations in practical deployments. In our discussion of single-layer defense, we analyze the various approaches that have been proposed to mitigate Byzantine faults at a single level of the network architecture. This includes cryptographic techniques, secure protocols, and intrusion detection systems that are tailored specifically for wireless networks. We evaluate their performance based on key metrics such as fault tolerance, scalability, and communication overhead. Furthermore, we also discuss the challenges and limitations of single-layer defense methods. We highlight the need for a more comprehensive and holistic approach that considers the multi-faceted nature of Byzantine faults. This leads us to explore multilayer defense methods, which combine techniques from different layers of the network architecture to provide a more robust and resilient defense against Byzantine faults. By providing a comprehensive introduction to the defense methods in Byzantine fault-tolerant consensus, this chapter aims to equip readers with the necessary knowledge and insights to design and deploy secure and reliable wireless networks in adversarial environments.

The single-layer defense method for Byzantine fault-tolerant consensus encompasses three critical layers: the physical layer, the protocol layer, and the data layer. Each layer contributes to a multi-faceted defense mechanism that ensures the security, reliability, and resilience of wireless networks. At the physical layer, defense strategies primarily focus on the protection and isolation of hardware devices. This includes measures such as physical access control, tamper-resistant hardware design, and secure hardware components. By strengthening the physical security of devices, we can mitigate the risk of malicious actors gaining unauthorized access or tampering with the hardware, which could potentially compromise the entire network. Moving to the protocol layer, defense mechanisms concentrate on ensuring the security and reliability of network communication. This involves the development and implementation of secure protocols that can detect and isolate faulty or malicious nodes. Techniques such as encryption, authentication, and secure routing are employed to ensure that messages are transmitted securely and that only authorized nodes can participate in the consensus process. Finally, at the data layer, defense strategies focus on protecting the integrity and consistency of data. This is crucial for maintaining the accuracy and trustworthiness of the consensus results. Data layer defense includes techniques such as data validation, error correction, and redundancy mechanisms. By employing these techniques, we can detect and correct errors or malicious modifications in the data, ensuring that the consensus process is based on accurate and reliable information. In summary, the single-layer defense method of Byzantine fault-tolerant consensus leverages defense strategies at the physicalclayer, protocol layer, and data layer to provide a comprehensive and multi-faceted defense against malicious or faulty actors in wireless networks. By strengthening each layer, we can significantly enhance the security, reliability, and resilience of the network.

3.3.1 Physical Layer Defense Techniques

Byzantine attacks on the physical layer mainly manifest as unstable network connections, such as signal tampering, signal blocking, and timed attacks. Many existing studies focus on designing consensus protocols for physical layer attacks to defend against a variety of physical layer attacks.

The methods to resist physical layer attacks include equipment reinforcement and physical isolation.

Equipment Reinforcement Enhance the security of hardware devices through physical means, such as reinforcing the equipment casing and installing monitoring cameras, to prevent damage caused by physical attacks on the equipment. Drawing from the communication failure model, Santoro et al. [18] conducted an in-depth study of the binary consensus problem, where nodes communicate and decide between two input values: 0 and 1. The objective of their research was to determine the threshold on the number of dynamic faults that would render achieving any meaningful form of agreement, including a strong majority, impossible. Furthermore, they sought to establish limitations on the number of dynamic faults under which meaningful consensus, including unanimity, could still be achieved. Notably, the constraints they derived are shown to be accurate across a wide variety of network topologies, such as hypercubes, toruses, rings, and complete graphs. This finding demonstrates the generalizability and practical relevance of their work. Additionally, the authors made a significant contribution by bridging the previous gap in understanding regarding the feasibility of achieving non-trivial consensus within complete graphs in the context of dynamic Byzantine faults. This achievement provides valuable insights into the design and operation of distributed systems under faulty conditions.

In the face of significant challenges posed by the inherent unreliability of wireless ad hoc networks, Moniz et al. [19] focused their research on overcoming the difficulty of achieving Byzantine fault-tolerant binary consensus within single-hop wireless ad hoc networks. Their groundbreaking work introduced Turquois, an asynchronous Byzantine consensus protocol specifically tailored to handle the dual threat posed by Byzantine nodes and dynamic omission transmission faults that often threaten to disrupt communication between reliable, functioning nodes. One of the standout features that contributes to Turquois's effectiveness is its robust ability to withstand dynamic message omissions. This crucial attribute plays a pivotal role in optimizing the utilization of the wireless broadcasting medium, ensuring that even in the presence of transmission errors, the network remains functional and can continue to achieve consensus; When considering Byzantine generals problems within a distributed system that aims for a desired level of fault tolerance, Lamport [45] outlined a generalized approach. Instead of relying solely on explicit timeouts, a simplified clock-driven algorithm was implemented, assuming reliable clock synchronization as well as a resolution to the Byzantine generals problem.

Acknowledging the potential for network failures in message delivery at the physical layer, Castro et al. [46] introduced a replication algorithm called Byzantine fault tolerance (BFT). This algorithm aims to enable the construction of highly available systems that can withstand Byzantine faults. The algorithm's recovery mechanism allows it to endure any number of faults throughout the system's lifecycle, provided that fewer than one-third of the replicas fail within a defined vulnerability window. To implement the first Byzantine fault-tolerant Network File System (NFS), the authors utilized the BFT implementation as a generic program library. The BFT library and Byzantine fault-tolerant system (BFS) exhibit excellent performance due to several crucial optimizations, chief among them the employment of symmetric cryptography for message authentication. The results of the performance evaluation reveal that BFS performs within a range of 2% faster to 24% slower compared to non-replicated production implementations of the NFS protocol.

Physical Isolation Isolate critical equipment from external networks to reduce the impact of physical layer attacks on the system. For example, using dedicated networks or physically isolated network switches to ensure communication security. Kim et al. [4] set out to tackle a particularly vexing challenge in vehicle-to-everything (V2X) networks: the peers' number tradeoff problem. This issue is exacerbated by the constant movement and shifting dynamics of nodes in such networks. To address this, they devised an innovative mechanism aimed at selecting an optimal subset of peers. The key to their approach lies in the utilization of reinforcement learning (RL) techniques. These techniques empower the network to adapt and respond effectively to the dynamic effects of node mobility, thus enhancing the resilience of the Fabric-enabled V2X network. Specifically, Kim et al. formulated the RL process as a contextual multiarmed bandit (MAB) problem. By framing the problem in this manner, they were able to leverage the power of reinforcement learning to learn from past experiences and make smarter decisions about peer selection. This not only helps to optimize the performance of the V2X network but also ensures that it remains robust and reliable, even in the face of constantly changing conditions; Xie et al. [47] recognized the significant challenge posed by large-scale concurrent access of wireless devices in completing consensus processes, as this could lead to overwhelming consumption of communication and computing resources. To address this, they introduced a novel Byzantine fault-tolerant (BFT) consensus protocol named AirCon, specifically tailored for wireless user equipment (UE) through the utilization of aerial computing. The key innovation of AirCon lies in its ability to leverage the inherent properties of wireless channels to achieve physical layer consistency automatically during reception from UEs, thereby significantly reducing the communication and computing overheads typically associated with traditional consistency protocols.

3.3.2 Protocol Layer Defense Techniques

Byzantine attacks represent a formidable challenge not only at the physical layer but also, critically, at the network protocol layer. The complexity of network protocols and their pivotal role in facilitating secure communication among nodes render them particularly vulnerable to such attacks. Therefore, significant research efforts have been directed toward developing Byzantine fault-tolerant consensus mechanisms specifically tailored to the protocol layer. The goal of these endeavors is to design robust protocols that can withstand the presence of malicious nodes and guarantee consistent agreement among the remaining honest nodes in the network. This requires the development of novel techniques and algorithms that can detect, isolate, and mitigate the impact of Byzantine faults at the protocol level. Such efforts often involve the integration of cryptographic primitives, such as encryption, authentication, and digital signatures, to ensure the authenticity, integrity, and confidentiality of messages exchanged over the network. Additionally, researchers explore the use of redundant communication paths, distributed ledger technologies, and reputation-based mechanisms to increase the resilience of the network against Byzantine attacks. By focusing on the protocol layer, these research efforts aim to establish secure communication channels and protocols that can safeguard the integrity of network operations and ensure that the consensus process remains robust and reliable in the presence of malicious actors. The development of Byzantine fault-tolerant consensus mechanisms at the protocol layer is crucial for building secure and resilient distributed systems that can withstand a wide range of malicious attacks.

The methods to resist network protocol layer attacks include encrypted communication, identity verification, and DoS attack defense.

Encrypted Communication Using encryption technology to encrypt communication data, ensuring the confidentiality and integrity of the data. This can prevent man-in-the-middle attacks and data tampering. Zhao et al. [48] introduced an innovative intergroup supervision mechanism into the multilayer practical Byzantine fault tolerance (PBFT) consensus process. This mechanism involves representative nodes from each consensus group verifying consensus messages within other groups, effectively deterring malicious primary nodes from disrupting intergroup consistency. A verifiable random function (VRF)-based group number allocation algorithm was also developed to ensure the smooth implementation of this supervision.

Furthermore, to maximize the scalability of the system model, an objective function was analyzed for optimizing the consensus structure. This function aims to identify a hierarchical grouping structure that minimizes communication complexity while maintaining a Byzantine fault tolerance of one-third. Lastly, a node reputation-based hierarchical strategy was devised to further bolster the security of the multilayer consensus model. Michael et al. [49] observed that the number of transmissions required to achieve consensus during the protocol stage of many-to-many communication still experiences a secondary increase. To address this, they conducted experiments demonstrating that, under practical settings, integrating random linear network coding (RLNC) with practical Byzantine fault tolerance (PBFT) into a hybrid protocol operating on a specific network layer significantly enhances the transmission efficiency of the secondary stage. This approach, when applied to other combinations of message-based consensus and network encoding protocols, not only boosts performance but also strengthens robustness and security.

Identity Verification Verify the identity of nodes participating in consensus to ensure that only legitimate nodes can participate in the consensus process. This can prevent illegal nodes from affecting the system. Saw et al. [50] introduced a counterfactual quantum Byzantine fault-tolerant (CQ-BFT) protocol tailored for multi-party networks. This protocol leverages counterfactual unitary remote computation coupled with chained quantum Zeno gates, enabling consensus among all parties without requiring physical particles to traverse the quantum channel. Highlighting the distinctive nature of counterfactual communication, they have shown that the CQ-BFT protocol functions without a shared phase reference, offering a quantum layer for unparalleled security and resilience against phase noise, thereby fulfilling the stringent demands of blockchain technology.

Buterin et al. [51] delved into the intricate problem of Byzantine behaviors emerging at the protocol layer within a trustless system. To counter this pressing challenge, they introduced the Casper protocol, a groundbreaking security-deposit-based Proof-of-Stake (PoS) mechanism tailored specifically for Ethereum v2.0. Ethereum, a renowned blockchain-based distributed computing platform and operating system, has long been at the forefront of blockchain innovation. Casper represents a hybrid consensus mechanism that bridges the gap between cutting-edge research in PoS algorithms and Byzantine fault-tolerant consensus theory. By incorporating these two distinct areas of expertise, Casper aims to ensure the integrity and resilience of the Ethereum network, even in the presence of malicious or faulty nodes exhibiting Byzantine behaviors; Within the context of UAV networks operating in highly complex environments, these networks effectively function as asynchronous Byzantine distributed systems. Such networks are comprised of lightweight nodes that are not only vulnerable to external physical interference and network-based attacks but also face the additional challenge of dynamically generated error nodes emerging within the network itself.

Given the inherent resource constraints that characterize UAV networks, Kong et al. [52] conducted a comprehensive study focused on the trustworthiness of the participating UAVs. The core of this study revolves around the timely identification and isolation of erroneous nodes within the network. To this end, Kong et al. introduced LAP-BFT, a lightweight and asynchronous provable Byzantine fault-tolerant consensus algorithm. LAP-BFT achieves a globally trusted assessment of the network state by first establishing consensus on the local status data of nodes within asynchronous, generically selected subsets.

DoS Attack Defense Using load balancing, traffic cleaning, and other technologies to defend against DoS attacks, ensuring that the system can operate normally and handle normal network traffic. Thakker et al. [53] explored the development of a novel consensus protocol, drawing inspiration from the practical Byzantine fault tolerance algorithm. This protocol aims to offer efficient and flexible applications to the Internet of Things (IoT), vital for enhancing scalability and reliability. However, the traditional PBFT algorithm lacks two crucial features: resilience against denial-of-service (DoS) attacks and high throughput for swift final consensus during attacks. To address these limitations, they analyzed the current PBFT consensus protocol implementation and proposed a novel security approach. This approach involves incorporating the characteristics of random hash generation and threshold comparison to bolster resistance against DoS attacks, thereby achieving consensus in a secure manner. Given that Byzantine fault-tolerant algorithms offer consistency in finality, they often come with drawbacks such as high communication overhead, vulnerability to denial-of-service attacks, limited scalability, and a low threshold for faulty nodes. Moreover, these algorithms are not tailored for IoT systems, which necessitate near-real-time transaction confirmation, maximum fault tolerance, and suitable transaction verification rules. To address these issues, Makhdoom et al. [54] introduced "pledge," an innovative consensus protocol grounded in honest proof, aiming to minimize the likelihood of malicious behavior in the blockchain consensus process.

3.3.3 Data Layer Defense Techniques

The complexities of Byzantine attacks at the data layer are also receiving significant attention from researchers. The data layer, being the foundation upon which distributed systems operate, holds the vital information that drives network operations and ensures the trustworthiness of the consensus process. As such, any malicious manipulation or corruption of data can have devastating consequences. To combat this, numerous investigations are being conducted to delve deeper into the mechanisms and patterns of Byzantine attacks at the data layer. These efforts aim to provide a comprehensive understanding of how attackers target and exploit vulnerabilities in the data layer to launch malicious attacks. By studying the attack vectors and techniques used by attackers, researchers can gain insights into how such attacks can be mitigated or prevented. To achieve this, researchers are exploring various techniques and methodologies. They are analyzing historical data and incident reports to identify patterns and trends in Byzantine attacks. They are also developing novel algorithms and tools to detect and prevent data corruption in distributed systems. In addition, they are exploring the use of secure data storage mechanisms, such as encryption and distributed ledgers, to protect the integrity and confidentiality of data. The ultimate goal of these investigations is to provide the distributed systems community with effective countermeasures and best practices for mitigating Byzantine attacks at the data layer. By doing so, we can ensure the integrity and security of data within distributed systems, building more resilient and trustworthy networks that can withstand the threats posed by malicious actors.

The methods to resist data layer attacks include data verification, data backup and recovery, and Byzantine fault-tolerant consensus algorithm.

Data Verification Verify the stored data to ensure its integrity and consistency. This can prevent Byzantine nodes from tampering or forging data. Chen et al.

[38] recognized and addressed several pressing issues with existing consensus mechanisms used in blockchain technology. The key challenges they identified were the excessive amount of information exchanged between network nodes and the straightforward approach to electing the master node. These issues not only hindered efficiency but also posed potential security risks. To address these shortcomings, Chen et al. introduced the Aggregate Threshold Ring Signature-Practical Byzantine Fault Tolerant (ATRS-PBFT) (ring signature-practical Byzantine fault-tolerant) consensus mechanism, which leverages the ATRS algorithm. This novel approach significantly reduces the number of required signatures and verifications, thereby boosting the overall efficiency of the consensus process. This efficiency gain is crucial in blockchain systems, where each transaction contributes to the overall load on the network. Moreover, ATRS-PBFT goes beyond mere efficiency improvements and addresses Byzantine faults occurring at the information layer; Fu et al. [55] observed that most current approaches rely on blockchain for cross-domain IoT data sharing. However, incremental IoT data sharing poses dual challenges in terms of consensus mechanism efficiency and Byzantine fault tolerance. Existing consensus mechanisms, when used alone, are unable to fulfill these demands concurrently. As a solution, they introduced an efficient and fault-tolerant blockchain consensus transformation (BCT) mechanism tailored specifically for the Internet of Things. Additionally, they developed two consensus algorithms—RAFT (DRAFT) and double-layer parallel Byzantine fault tolerance (DPBFT)—to enhance the efficiency and fault tolerance of the data-sharing process.

Data Backup and Recovery Establish a data backup and recovery mechanism to ensure rapid recovery of data in the event of data loss or damage. This can reduce the impact of data layer attacks on the system. The conventional consensus mechanisms relied heavily on computational resources to achieve distributed consensus, leaving them vulnerable to sybil and double-spending attacks from malicious nodes. Moreover, the frequency of such attacks on blockchain systems has been escalating. To address these challenges, Xu et al. [56] introduced the Fast Resilient tamper-proof blockchain (FRTB), a novel blockchain structure built upon block data and a data-protected storage file allocation table (FAT). This architecture, an enhancement of the traditional blockchain framework, incorporates a block data protection system and block data storage file allocation table information, significantly impacting the blockchain's traceability and tamper resistance. The FRTB boasts features like physical tamper resistance, enhanced traceability, and rapid distributed consensus.

Byzantine Fault-Tolerant Consensus Algorithm Byzantine fault-tolerant consensus algorithms such as practical Byzantine fault tolerance (PBFT) are used to achieve fault tolerance and consensus at the data layer. These algorithms can ensure system availability and security in the presence of Byzantine errors. Chen et al. [57] presented a reputation-driven, two-tier enhanced practical Byzantine fault-tolerant consensus computing framework called Dual Trust-Practical Byzantine Fault Tolerant (DT-PBFT), aimed at implementing PBFT in expansive IoT ecosystems and blockchain systems. To facilitate seamless integration with large-scale IoT setups, they assign attributes like regional positioning, communication capabilities, and computational prowess to each participating node. This allows client nodes to leverage these attributes as inputs for reputation-based algorithms to select master nodes for each tier in the two-tier consensus model. This reputation-based master node selection enhances the trustworthiness of the consensus algorithm. Furthermore, DT-PBFT eliminates the preliminary preparation phase and thus reduces communication complexity. Hao et al. [58] highlighted a crucial gap in current private blockchain platforms, which lack adequate theoretical and data-driven insights for analyzing the efficacy of consensus algorithms. To address this, they introduced an evaluation framework that assesses the performance of consensus algorithms on Ethereum's and Hyperledger Fabric's private blockchain implementations. By conducting a quantitative analysis focusing on latency and throughput, they derived performance metrics for various consensus algorithms under varying transaction volumes. Their findings revealed that the consistency mechanism poses significant performance challenges. Notably, the study showed that, regardless of the workload, practical Byzantine fault tolerance demonstrated superior performance in terms of latency and throughput compared to Proof-of-Work (PoW). Pan et al. [59] introduced a hybrid consensus algorithm tailored for node groups exhibiting hierarchical structures. Their approach combines the strengths of the practical Byzantine fault tolerance algorithm and the RAFT algorithm, while addressing their respective limitations, to better suit the characteristics of hierarchical systems. Initially, each node is assigned a permission value based on its hierarchical position, and the nodes are then organized into layers according to these values. The lower layers employ the RAFT algorithm, while the upper layers utilize the PBFT algorithm. This hierarchical arrangement significantly reduces communication overhead and efficiently enhances consensus performance. Furthermore, they introduced an authoritative value allocation strategy within PBFT to effectively mitigate malicious attacks. In the RAFT algorithm, the inclusion of monitoring nodes enhances its Byzantine fault tolerance.

3.3.4 *Cross-Layer Defense Techniques*

The multilayered defense method in Byzantine fault-tolerant consensus (BFT) is a sophisticated approach that endeavors to guarantee the consistency and security of state consensus among non-Byzantine nodes (normal nodes) in distributed systems infiltrated with potentially faulty or malicious Byzantine nodes. This method goes beyond a single layer of defense and employs a holistic, collaborative approach that leverages the defenses of the physical layer, network protocol layer, and data layer to address a wide range of challenges. These challenges include message loss, message disorder, delayed message arrival, and any malicious behavior that Byzantine nodes may exhibit, including collusion. The method's primary objective is to ensure that non-Byzantine nodes can operate unperturbed by the Byzantine nodes, utilizing the same input information to generate identical results, thus maintaining the system's security and flexibility. This multilayered defense approach is specifically tailored to achieve efficient and reliable consensus mechanisms in complex network environments, especially in decentralized systems such as blockchain. The reason for this is that decentralized systems often operate in environments where Byzantine nodes are a significant concern, and traditional consensus mechanisms may not be sufficient to ensure the system's resilience and reliability.

Recognizing that complex attacks often require policy convergence across multiple layers to succeed, research efforts have increasingly focused on investigating cross-layer Byzantine scenarios in wireless networks. These comprehensive surveys aim to provide a deeper understanding of how Byzantine attacks propagate across different layers and develop targeted strategies to mitigate their impact. By gaining insights into the interplay between different layers, researchers can design more effective defense mechanisms that can thwart attacks at multiple levels, further enhancing the security and reliability of distributed systems.

In anticipation of the dense and large-scale wireless network scenarios that will likely be commonplace in future Internet of Things (IoT) deployments, Jiang et al. [24] zeroed in on the formidable challenge of achieving low-delay and high-throughput Byzantine fault-tolerant (BFT) consensus in permissionless systems. Alongside this, they also tackled the vulnerability to sybil attacks, a significant threat in such networks. To address these challenges, Jiang et al. introduced SENATE, a sybil-resistant wireless network coordinate-based Byzantine consensus mechanism. SENATE is an innovative approach that boasts the capability to achieve real-time consensus, even within networks comprising hundreds of nodes, with consensus delays of less than one second. This is achieved by capitalizing on the fact that even a compromised or faulty node cannot convincingly manipulate its wireless channel interactions with other nodes. SENATE leverages unique wireless fingerprints, such as channel state information or received signal strength, to distinguish and identify nodes. This approach significantly enhances the security and trustworthiness of the consensus process, making it resilient to sybil attacks and other malicious behaviors.

Leng et al. [44] delved deeply into the intricate complexities that often accompany practical Byzantine fault-tolerant consensus algorithms. These algorithms frequently struggle with issues like soaring communication overhead, compromised security, scalability limitations, and the challenges of maintaining traceability. To tackle these multifaceted challenges effectively, the researchers introduced a groundbreaking Byzantine fault-tolerant consensus algorithm called the generalized practical Byzantine fault-tolerant consensus algorithm (GPBFT). The GPBFT algorithm is built upon a robust framework that leverages the concept of dual administrators and short group signatures. This framework is specifically designed to ensure that the algorithm can be practically implemented across a wide range of domains, including e-commerce, e-banking, e-voting, and e-auction systems. By addressing the key challenges faced by traditional Byzantine fault-tolerant consensus algorithms, GPBFT demonstrates its versatility and applicability in various real-world scenarios, ensuring the reliability, security, and scalability of consensus processes across multiple industries.

As an example, Gilad et al. [60] delved deeply into the challenges posed by Eclipse attacks and sybil attacks in the context of blockchain technology.

In response, they introduced a novel cryptocurrency caving confirmation times on the order of one minute. Algorand was also designed to scale effectively, accommodating a vast user base while maintaining its efficiency. To combat the potential threat of sybil attacks, where malicious actors attempt to flood the network with fraudulent identities, Algorand employs a unique mechanism that assigns a weight to each user. This approach effectively mitigates the risk of such attacks, ensuring the network remains resilient to such malicious behavior. The cornerstone of Algorand's consensus mechanism lies in a Byzantine agreement protocol that utilizes verifiable random functions (VRF). This protocol allows Algorand to efficiently and promptly reach a consensus among network users regarding the next set of transactions and the creation of a new block. This consensus is achieved with low latency and eliminates the possibility of forks in the blockchain, thus significantly enhancing the security and robustness of the entire system.

Within the realm of mobile networks, where faults and agent mobility are not merely occasional occurrences but rather commonplace and often inherent to the environment, Zou et al. [61] dedicated their research to tackling the fault-tolerant consensus problem. In such networks, faulty agents can exhibit more than just crash failures; they may also attempt to disseminate incorrect information, posing a significant threat to the integrity of the consensus process. To bolster the resilience of consensus in these challenging environments, Zou et al. ingeniously adopted the non-orthogonal multiple access (NOMA) technique from 5G technology. NOMA enables efficient data transmissions between agents in mobile networks, ensuring that vital information flows smoothly and without delay. Furthermore, they proposed a distributed algorithm tailored specifically to achieve fault-tolerant consensus in mobile networks that are vulnerable to Byzantine faults. This algorithm is crafted to operate effectively in scenarios where faulty agents can exhibit malicious behavior, such as propagating incorrect information. By incorporating mechanisms to detect and mitigate the influence of faulty agents, the algorithm ensures the reliability and accuracy of the consensus process, even in the face of potential malicious actors.

In the context of blockchain-based Byzantine fault-tolerant consensus, Wang et al. [62] conducted a comprehensive study of Byzantine failures occurring at both the protocol layer and the information layer. Their work culminated in the development of a cutting-edge consensus protocol known as the adaptive Byzantine fault-tolerant (ABFT) consensus protocol. ABFT incorporates an optimistic response assumption, a strategy that assumes honest behavior until proven otherwise. This approach, coupled with an adaptive mechanism for consensus formation, allows the protocol to achieve consensus with a latency of 2Δ in stable conditions. A key advantage of ABFT lies in its ability to tolerate up to half of the Byzantine failures within the network. This remarkable robustness ensures that the security of the system is upheld even in scenarios where synchronization among network participants is not stringently enforced. By enhancing the reliability and resilience of the system, ABFT paves the way for more secure and efficient blockchain-based applications.

Acknowledging the potential threats posed by the 51% attack, sybil attack, double-spending attack, and jamming attack, Zhang et al. [17] embarked on a research journey to devise a Byzantine fault-tolerant consensus protocol tailored specifically for single-hop wireless networks, where challenges like signal-to-noise ratio constraints are paramount. The resulting protocol, dubbed low-latency and reliable Byzantine fault-tolerant consensus protocol (LRBP), was designed to satisfy two properties: persistence and liveness. Persistence, a cornerstone of blockchain security, ensures that once a transaction is validated and recorded in the blockchain ledger, it becomes immutable, meaning it cannot be undone or invalidated by malicious actors. This property is critical in maintaining the integrity and authenticity of the blockchain data. Liveness, on the other hand, guarantees that the network will inevitably reach consensus on transactions, regardless of the presence of Byzantine failures or network disruptions. Even in challenging scenarios where malicious nodes or external interference attempts to hinder the consensus process, LRBP ensures that the network remains operational and able to finalize transactions. By incorporating these two fundamental properties, LRBP offers a robust and secure consensus mechanism for single-hop wireless networks, enabling reliable and efficient blockchain-based applications in environments with stringent constraints.

Jing et al. [63] made groundbreaking advancements in tackling a particularly vexing challenge: attaining majority consensus in environments where Byzantine fault-prone edge devices operate, while also contending with the unpredictable nature of an unreliable wireless channel. Their work stands out as truly pioneering, as it comprehensively considers the potential malicious faults that can originate from Byzantine devices at various system layers, spanning the physical layer, protocol layer, and information layer. To address this multifaceted problem, Jing et al. defined a novel concept of an (a, b)-majority consensus problem. This problem formulation encapsulates the need to achieve consensus with a specific threshold b, even in the presence of Byzantine devices that can disrupt the consensus process. To realize this goal, they devised a distributed protocol that ensures b-majority agreement within $\Theta(n)$ time steps, with a high probability of success. By developing such a comprehensive solution, Jing et al. have not only pushed the boundaries of Byzantine fault-tolerant consensus protocols but also laid the foundation for more secure and reliable distributed systems, especially in environments with unreliable wireless channels and vulnerable edge devices.

Homoliak et al. [64] focused their research on the systematic organization and broadening of understanding pertaining to blockchain security and privacy, significantly contributing to standardization efforts in this domain. To this end, we introduce the Security Reference Architecture (SRA) for blockchain, employing a layered model analogous to the ISO/OSI stack. This framework delineates the attributes and hierarchical structure of various security and privacy considerations. The SRA comprises four distinct layers: (1) the network layer, (2) the consensus layer, (3) the replication state machine layer, and (4) the application layer. Within each layer, we identify prevalent security threats, their sources, and corresponding countermeasures, while also examining the intricate dependencies across different layers.

3.4 Summary

This chapter provides a detailed introduction to the Byzantine fault-tolerant consensus mechanism. First, it introduces that the Byzantine fault-tolerant system protocol mainly includes multiple aspects, such as network, scheduling and consensus, and emphasizes the dominant position of Byzantine consensus. Secondly, common attack methods in Byzantine fault-tolerant consensus were introduced at the physical layer, protocol layer, and data layer. Then, for the three levels of attack methods, this chapter lists the corresponding defense methods in the Byzantine fault-tolerant consensus. Finally, examples were given to illustrate the application of Byzantine fault-tolerant consensus mechanisms in specific application scenarios such as the Internet of Things, blockchain, vehicle networks, and drone networks. We hope that the wireless network fault-tolerant consensus mechanism can provide inspiration for researchers to develop more practical consensus protocols.

References

1. Xiao, B., Lao, L., Dai, X., & Guo, S. (2020). G-pbft: a location-based and scalable consensus protocol for iot-blockchain applications. In *2020 IEEE International Parallel and Distributed Processing Symposium (IPDPS)* (pp. 664–673). IEEE.
2. Li, E., Feng, C. F., & Zhang, L. (2020). A scalable multi-layer pbft consensus for blockchain. *IEEE Transactions on Parallel and Distributed Systems, 32*(5), 1146–1160.
3. Gao, N., Huo, R., Wang, S., & Huang, T. (2023). Fibft: An improved byzantine consensus mechanism for edge computing. In *2023 IEEE Wireless Communications and Networking Conference (WCNC)* (pp. 1–6).
4. Kim, S., & Ibrahim, A. S. (2023). Byzantine-fault-tolerant consensus via reinforcement learning for permissioned blockchain-empowered v2x network. *IEEE Transactions on Intelligent Vehicles, 8*(1), 172–183.
5. Li, T., Tseng, L., Higuchi, T., Ucar, S., & Altintas, O. (2021). Poster: Fault-tolerant consensus for connected vehicles: A case study. In *2021 IEEE Vehicular Networking Conference (VNC)* (pp. 133–134).
6. Tseng, T.-Y., Huang, D.-J., Lin, J.-Y., Chang, P.-J., Lin, C.-W., & Liu, C. (2023). Consensus-based fault-tolerant platooning for connected and autonomous vehicles. In *2023 IEEE Intelligent Vehicles Symposium (IV)* (pp. 1–8).
7. Lingjun Kong, B. C., & Hu, F. (2022). Lap-bft: lightweight asynchronous provable byzantine fault-tolerant consensus mechanism for uav network trusted systems. In *Web and Big Data - 6th International Joint Conference, APWeb-WAIM 2022* (pp. 232–246).
8. Zhang, M., Dong, C., & Huang, Y. (2019). Fs-mac: An adaptive mac protocol with fault-tolerant synchronous switching for fanets. *IEEE Access, 7*, 80602–80613.
9. Zhang, L., Liao, Z., & Dong, Z. (2021). Uav swarm exploration with byzantine fault tolerance. In *2021 China Automation Congress (CAC)* (pp. 7150–7154).
10. Lamport, L., Shostak, R. E., & Pease, M. C. (1982). The byzantine generals problem. *ACM Transactions on Programming Languages and Systems, 4*(3), 382–401.
11. DeGroot, M. H. (1974). Reaching a consensus. *Taylor & Francis Journal of the American Statistical association, 69*(345), 118–121.
12. Dolev, D., Fischer, M. J., Fowler, R. J., Lynch, N. A., & Strong, H. R. (1982). An efficient algorithm for byzantine agreement without authentication. *Information and Control, 52*(3), 257–274.

13. Fischer, M. J., & Lynch, N. A. (1982). A lower bound for the time to assure interactive consistency. *Information Processing Letters, 14*(4), 183–186.
14. Zhou, S., & Ying, B. (2021). Vg-raft: An improved byzantine fault tolerant algorithm based on raft algorithm. In *2021 IEEE 21st International Conference on Communication Technology (ICCT)* (pp. 882–886).
15. Wu, J., Su, M., Xu, X., Qiao, L., Dai, M., & Cao, W. (2024). Less sample-cooperative spectrum sensing in the presence of large-scale byzantine attack. *IEEE Sensors Letters, 8*(1), 1–4.
16. Wissner-Gross, A. D., Willard, J. C., & Weston, N. (2020). Tamper-proofing imagery from distributed sensors using learned blockchain consensus. In *2020 IEEE Applied Imagery Pattern Recognition Workshop (AIPR)* (pp. 1–4).
17. Zhang, L., Zhang, B., & Li, C. (2023). An efficient and reliable byzantine fault tolerant blockchain consensus protocol for single-hop wireless networks. *IEEE Transactions on Wireless Communications, 23*(3), 1974–1987.
18. Soesanto, D., Adji, T. B., & Ardiyanto, I. (2022). Adaptive proof of work architecture design by implementing multiple mempool. In *2022 International Conference on Advanced Creative Networks and Intelligent Systems (ICACNIS)* (pp. 1–8).
19. Moniz, H., Neves, N. F., & Correia, M. (2013). Byzantine fault-tolerant consensus in wireless ad hoc networks. *IEEE Transactions on Mobile Computing, 12*(12), 2441–2454.
20. Lei, K., Zhang, Q., Xu, L., & Qi, Z. (2018). Reputation-based byzantine fault-tolerance for consortium blockchain. In *2018 IEEE 24th International Conference on Parallel and Distributed Systems (ICPADS)* (pp. 604–611).
21. Melih Tas, I., & Baktir, S. (2024). Blockchain-based caller-id authentication (bbca): A novel solution to prevent spoofing attacks in voip/sip networks. *IEEE Access, 12*, 60123–60137.
22. Nagananda, K. G., Blum, R. S., & Koppel, A. (2020). Reduced-rank least squares parameter estimation in the presence of byzantine sensors. In *2020 54th Annual Conference on Information Sciences and Systems (CISS)* (pp. 1–6).
23. Santoro, N., & Widmayer, P. (October 2007). Agreement in synchronous networks with ubiquitous faults. *Theoretical Computer Science, 384*(2), 232–249.
24. Jiang, Z., Cao, Z., Krishnamachari, B., Zhou, S., & Niu, Z. (2020). Senate: A permissionless byzantine consensus protocol in wireless networks for real-time internet-of-things applications. *IEEE Internet of Things Journal, 7*(7), 6576–6588.
25. Thawani, S., & Upadhyay, H. (2015). Securing tora against sybil attack in manets. In *2015 International Conference on Futuristic Trends on Computational Analysis and Knowledge Management (ABLAZE)* (pp. 475–478).
26. Tsikerdekis, M. (2017). Identity deception prevention using common contribution network data. *IEEE Transactions on Information Forensics and Security, 12*(1), 188–199.
27. Tan, Y., Hong, K., Sengupta, S., & Subbalakshmi, K. P. (2011). Using sybil identities for primary user emulation and byzantine attacks in dsa networks. In *2011 IEEE Global Telecommunications Conference - GLOBECOM 2011* (pp. 1–5).
28. Rabieh, K., Mahmoud, M. M. E. A., Guo, T. N., & Younis, M. (2015). Cross-layer scheme for detecting large-scale colluding sybil attack in vanets. In *2015 IEEE International Conference on Communications (ICC)* (pp. 7298–7303).
29. Mrabet, Z. E., Selvaraj, D. F., Nair, A. S., & Ranganathan, P. (2019). Detection of the false data injection attack in home area networks using ann. In *2019 IEEE International Conference on Electro Information Technology (EIT)* (pp. 176–181).
30. Dai, J., Yang, J., Wang, Y., & Xu, Y. (2023). Blockchain-enabled cyber-resilience enhancement framework of microgrid distributed secondary control against false data injection attacks. *IEEE Transactions on Smart Grid, 15*(2), 2226–2236.
31. Okegbile, S. D., Cai, J., Chen, J., & Yi, C. (2024). A reputation-enhanced shard-based byzantine fault-tolerant scheme for secure data sharing in zero trust human digital twin systems. *IEEE Internet of Things Journal, 11*(12), 22726–22741.
32. Efendy, R. A., Almaarif, A., Budiono, A., Saputra, M., Puspitasari, W., & Sutoyo, E. (2019). Exploring the possibility of usb based fork bomb attack on windows environment. In *2019 International Conference on ICT for Smart Society (ICISS)* (vol. 7, pp. 1–4).

33. Wang, K., Wang, Y., & Ji, Z. (2020). Defending blockchain forking attack by delaying mtc confirmation. *IEEE Access, 8*, 113847–113859.
34. Brandenburger, M., Cachin, C., Lorenz, M., & Kapitza, R. (2017). Rollback and forking detection for trusted execution environments using lightweight collective memory. In *2017 47th Annual IEEE/IFIP International Conference on Dependable Systems and Networks (DSN)* (pp. 157–168).
35. Borzdov, B., Minchenok, M., & Yanovich, Y. (2023). Practical vulnerabilities in byzantine fault-tolerant blockchain consensus protocols. In *2023 XVIII International Symposium Problems of Redundancy in Information and Control Systems (REDUNDANCY)* (pp. 94–99). IEEE.
36. Zebari, R. R., Zeebaree, S. R. M., & Jacksi, K. (2018). Impact analysis of http and syn flood ddos attacks on apache 2 and iis 10.0 web servers. In *2018 International Conference on Advanced Science and Engineering (ICOASE)* (pp. 156–161).
37. Biagioni, E. (2019). Preventing udp flooding amplification attacks with weak authentication. In *2019 International Conference on Computing, Networking and Communications (ICNC)* (pp. 78–82).
38. Chen, L., Wang, C., & Jiang, W. (2023). An efficient byzantine fault-tolerant consensus mechanism based on aggregate threshold ring signature. In *2023 8th International Conference on Computer and Communication Systems (ICCCS)* (pp. 430–439).
39. Shaofeng, L., Yang, Z., Yao, Z., & Lin, N. (2023). Research on anti-leakage and tampering guarantee strategy for experimental data sharing based on blockchain. In *2023 IEEE 3rd International Conference on Electronic Technology, Communication and Information (ICETCI)* (pp. 277–281).
40. Lin, B., Xu, F., Xu, S., Wu, N., & Liu, Z. (2021). Lrpbft-algorithm-based consortium blockchain traceability scheme of dangerous goods in commercial ports. In *2021 13th International Conference on Wireless Communications and Signal Processing (WCSP)* (pp. 1–6). IEEE.
41. Wang, Y., Tian, M., Zhang, Y., Liu, X., Xiao, Y., & Hei, X. (2022). A security enhancement scheme for raft consensus algorithm against term forgery attacks. In *2022 IEEE 21st International Conference on Ubiquitous Computing and Communications (IUCC/CIT/DSCI/SmartCNS)* (pp. 175–183).
42. Katole, R. A., Sherekar, S. S., & Thakare, V. M. (2018). Detection of sql injection attacks by removing the parameter values of sql query. In *2018 2nd International Conference on Inventive Systems and Control (ICISC)* (pp. 736–741).
43. Ravi, N., Scaglione, A., & Nedić, A. (2019). A case of distributed optimization in adversarial environment. In *ICASSP 2019 - 2019 IEEE International Conference on Acoustics, Speech and Signal Processing (ICASSP)* (pp. 5252–5256).
44. Yu, X., Qin, J., Chen, P., & Leng, J. (2022). Gpbft: A practical byzantine fault-tolerant consensus algorithm based on dual administrator short group signatures. *Security and Communication Networks*, 2022.
45. Lamport, L. (1984). Using time instead of timeout for fault-tolerant distributed systems. *ACM Transactions on Programming Languages and Systems, 6*(2), 254–280.
46. Castro, M., & Liskov, B. (2002). Practical byzantine fault tolerance and proactive recovery. *ACM Transactions on Computer Systems, 20*(4), 398–461.
47. Xie, X., Hua, C., Hong, J., Gu, P., & Xu, W. (2023). Aircon: Over-the-air consensus for wireless blockchain networks. *IEEE Transactions on Mobile Computing, 23*(5), 4566–4582.
48. Zhao, Y., Guo, B., Qin, C., & Zhao, M. (2022). A multi-layer pbft consensus algorithm with inter-group supervision. In *2022 IEEE 24th Int Conf on High Performance Computing & Communications; 8th Int Conf on Data Science & Systems; 20th Int Conf on Smart City; 8th Int Conf on Dependability in Sensor, Cloud & Big Data Systems & Application (HPCC/DSS/SmartCity/DependSys)* (pp. 1101–1108). IEEE.
49. Braun, M., Wiesmaier, A., Alnahawi, N., & Geißler, J. (2021). On message-based consensus and network coding. In *2021 12th International Conference on Network of the Future (NoF)* (pp. 1–9). IEEE.

50. Paing, S. N., Setiawan, J. W., Ullah, M. A., Zaman, F., Duong, T. Q., Dobre, O. A., & Shin, H. (2024). Counterfactual quantum byzantine consensus for human-centric metaverse. *IEEE Journal on Selected Areas in Communications, 42*(4), 905–918.
51. Buterin, V., & Griffith, V. (2017). Casper the friendly finality gadget. *CoRR*, abs/1710.09437.
52. Kong, L., Chen, B., & Hu, F. (2023). Lap-bft: Lightweight asynchronous provable byzantine fault-tolerant consensus mechanism for uav network trusted systems. In B. Li, L. Yue, C. Tao, X. Han, D. Calvanese, & T. Amagasa (Eds.), *Web and big data* (pp. 232–246). Cham: Springer Nature Switzerland.
53. Thakker, J., & Park, Y. (2020). Resilient and efficient blockchain consensus protocol for internet-of-things. In *2020 IEEE International Conference on Consumer Electronics (ICCE)* (pp. 1–6).
54. Makhdoom, I., Tofigh, F., Zhou, I., Abolhasan, M., & Lipman, J. (2020). Pledge: A proof-of-honesty based consensus protocol for blockchain-based iot systems. In *2020 IEEE International Conference on Blockchain and Cryptocurrency (ICBC)* (pp. 1–3).
55. Fu, J., Zhang, L., Wang, L., & Li, F. (2021). Bct: an efficient and fault tolerance blockchain consensus transform mechanism for iot. *IEEE Internet of Things Journal, 10*(14), 12055–12065.
56. Xu, Y., Wang, Y., & Li, T. (2023). Frtb-fast blockchain architecture based on block data and data protection system's file partition tables. In *2023 IEEE 3rd International Conference on Power, Electronics and Computer Applications (ICPECA)* (pp. 136–140).
57. Chen, Y., & Jia, Y. (2023). Dt-pbft: A double-layer group consensus algorithm of credibility for iot blockchain. In *2023 2nd International Conference on Big Data, Information and Computer Network (BDICN)* (pp. 292–299). IEEE.
58. Hao, Y., Li, Y., Dong, X., Fang, L., & Chen, P. (2018). Performance analysis of consensus algorithm in private blockchain. In *2018 IEEE Intelligent Vehicles Symposium (IV)* (pp. 280–285). IEEE.
59. Pan, Y., Wei, W., Liu, Y., & Zhu, C. (2021). Hybrid consensus algorithm based on hierarchy authority. In *2021 IEEE International Conference on Electronic Technology, Communication and Information (ICETCI)* (pp. 553–559). IEEE.
60. Gilad, Y., Hemo, R., Micali, S., Vlachos, G., & Zeldovich, N. (2017). Algorand: Scaling byzantine agreements for cryptocurrencies. In *Proceedings of the 26th Symposium on Operating Systems Principles*, SOSP '17 (pp. 51–68), New York, NY, USA. Association for Computing Machinery.
61. Zou, Y., Xu, M., Yu, J., Zhao, F., & Cheng, X. (2022). Fault-tolerant consensus with noma in mobile networks. *IEEE Wireless Communications, 29*(3), 80–86.
62. Wang, R., Tsai, W.-T., Zhang, F., Yu, L., Zhang, H., & Zhang, Y. (2022). Adaptive byzantine fault-tolerant consensusprotocol. In M. Qiu, Z. Lu, & C. Zhang, (Eds.), *Smart Computing and Communication - 7th International Conference, SmartCom 2022, New York City, NY, USA, November 18–20, 2022, Proceedings*, vol. 13828 of *Lecture Notes in Computer Science* (pp. 66–75). Springer.
63. Jing, G., Zou, Y., Yu, D., Luo, C., & Cheng, X. (2023). Efficient fault-tolerant consensus for collaborative services in edge computing. *IEEE Transactions on Computers, 72*(8), 2139–2150.
64. Homoliak, I., Venugopalan, S., Reijsbergen, D., Hum, Q., Schumi, R., & Szalachowski, P. (2021). The security reference architecture for blockchains: Toward a standardized model for studying vulnerabilities, threats, and defenses. *IEEE Communications Surveys & Tutorials, 23*(1), 341–390.

Chapter 4
Wireless Blockchain

Wireless blockchain protocols are an innovative technology that applies blockchain technology to the wireless network domain. They can be categorized into two main approaches. The first approach involves directly deploying blockchain protocols into wireless networks. This method utilizes existing blockchain technologies, such as Bitcoin or Ethereum, and applies them to wireless networks to achieve decentralized data sharing. By doing so, blockchain can provide an immutable distributed ledger, recording the communications and transactions of nodes within the wireless network, thereby enhancing the network's security and transparency. However, this direct deployment approach may face challenges in practical applications, such as the high computational and storage demands of blockchain potentially conflicting with the resource limitations of wireless devices. The second approach focuses on enhancing the performance of wireless blockchain. To overcome the resource constraints and environmental complexities of wireless networks, researchers have proposed various optimization solutions. These include improving consensus algorithms to reduce computational and communication overhead, adopting lightweight blockchain frameworks to suit resource-constrained wireless devices, and developing specialized resource management and allocation mechanisms to optimize network performance. These methods aim to increase the efficiency and scalability of blockchain in wireless networks, ensuring that efficient and reliable blockchain operations can be achieved in wireless environments.

This chapter delves into two key aspects of wireless blockchain protocols: the basic concept of blockchain protocols and wireless blockchain protocols.

- Section 4.1 introduces the basic concepts of blockchain protocols and their requirements for the network layer.
- Section 4.2 discusses two application methods of wireless blockchain protocols and presents two blockchain protocol designs: BLOWN and wChain.

M. Xu et al., *Wireless Consensus*, Wireless Networks,
https://doi.org/10.1007/978-3-031-70859-6_4

4.1 Introduction to Blockchain

4.1.1 Overview of Blockchain

Figure 4.1 depicts a prototypical model of a blockchain, which abstracts the core structure of Bitcoin. This model represents a blockchain as a decentralized ledger configured as a sequential chain of blocks, each linked via cryptographic hashes and distributed across the nodes of a peer-to-peer network. The blockchain initiates with a "genesis block," also referred to as block zero. A full node within the blockchain network retains a complete copy of the blockchain stored in a local database. In contrast, a light node maintains only the headers of blocks, enabling it to verify block authenticity through simplified payment verification (SPV). This configuration, however, exposes light nodes to security vulnerabilities such as potential double-spending attacks due to their reliance on an incomplete version of the blockchain. Delving deeper into the blockchain's structure, each block is constituted by a header and a body. Every block, with the exception of the genesis block, connects to its predecessor by referencing the predecessor's hash. Typically, the header includes a timestamp, a nonce, a previous hash, a target, and a Merkle root, whereas the body comprises serialized transactions.

Furthermore, three key components illustrated in Fig. 4.1 require additional elucidation: the hash function, the Merkle tree, and the nonce.

Similar to the way fingerprints uniquely identify a person, a hash function is engineered to map any length of data to a specific, unchanging value, denoted as $h := H(x) : \{1, 0\}^* \to \{1, 0\}^l$.

In a manner similar to fingerprinting, hash functions are engineered to assign data of arbitrary length to a specific, fixed value, expressed as $h := H(x) : \{1, 0\}^* \to \{1, 0\}^l$. The family of Secure Hash Algorithms (SHA)[1] is standardized by the National Institute of Standards and Technology (NIST) and is recognized for

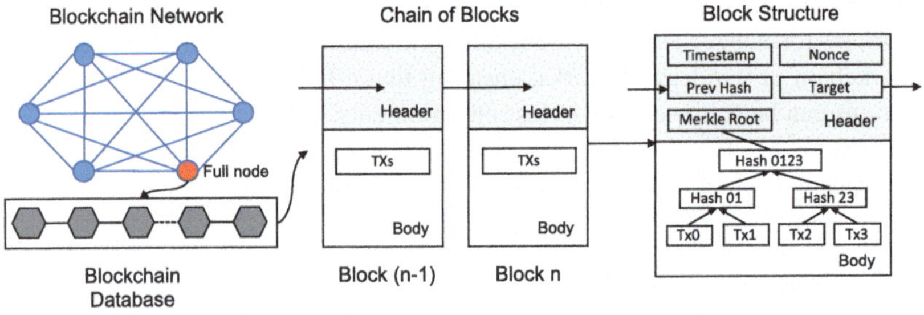

Fig. 4.1 A blockchain prototype

[1] The predominant hash function in blockchain technology is SHA-256, which is deemed secure for the present, even though polynomial-time quantum algorithms for factoring and discrete logarithms have theoretically exposed potential security threats to existing hash functions.

its collision resistance, making it challenging for any probabilistic polynomial-time (PPT) algorithm to find two distinct inputs, x_1 and x_2, that satisfy $H(x_1) = H(x_2)$ [1]. Like a fingerprint, h allows for convenient transfer (fixed and limited length) and secure verification ($H(x)$ being a one-way function).

The Merkle tree, first proposed by Ralph Merkle [2], is utilized for organizing multiple cryptographic hashes within a binary tree structure. As depicted in Fig. 4.1, inputs populate the leaves of the Merkle tree, while each parent node is formed by hashing the values of its child nodes. Employing a Merkle tree enables a block to merely contain a solitary Merkle root in its header. Furthermore, the computational expense required for verifying an individual transaction is $O(\log N)$, where N represents the total count of inputs [3].

In the realm of Bitcoin, a cryptographic nonce is employed as a one-time, uniquely generated random number. Miners use this nonce to validate a Proof-of-Work (PoW) by discovering a value such that the block header's double hash falls below a specified target. Conversely, in different systems such as Ethereum, the term "nonce" adopts a distinct meaning, serving as a transaction counter in each account to thwart double-spending attempts.

Within cryptocurrency systems, it is crucial that transactions, that represent monetary transfers, undergo verification prior to their integration into a block. The model governing a transaction significantly influences its structure, the validation processes it must pass through, and how balances are calculated. Presently, the two predominant models are the unspent transaction output (UTXO) model and the account/balance model. In the UTXO model, transactions utilize funds from previously received UTXOs and generate new UTXOs for future expenditures, necessitating clear specification of ownership and value for verification. A node's balance under this model is the sum of all UTXOs it controls. Alternatively, the account/balance model operates akin to conventional banking systems where each account maintains a balance, and transactions are processed by debiting one account and crediting another. Each model presents its own set of strengths and weaknesses concerning security, scalability, simplicity, operational efficiency, and storage demands.

- *Security:* Within the UTXO framework, each transaction must incorporate legitimate UTXOs, and the aggregate value of these UTXOs should surpass the combined amount of the transaction outputs. To prevent the risk of double-spending, UTXOs are designed to be spent only once. Conversely, in the account/balance model, an elementary prevention technique against double-spending is to integrate a sequential counter in every transaction. When presented with two conflicting transactions, validators opt for the transaction that displays a lower sequence counter. Furthermore, the integration of a unique chain identity and nonce in both models aids in preventing replay attacks.
- *Scalability:* The UTXO model facilitates the parallel execution of transactions because each UTXO operates independently, allowing disparate UTXOs to be expended in simultaneous transactions. In the account/balance model, however, transactions are processed in a sequential manner to maintain account integrity.

- *Efficiency:* Though inherently complex, the UTXO model can lead to inefficiencies within blockchain systems if not executed correctly. Enhancing the UTXO model's efficiency involves strategies such as maintaining a compact version of the UTXO set by each full node, which represents only a fraction of the total UTXOs within the network. This subset is dynamically updated with each transaction, simplifying the validation process to merely confirming the legality of involved UTXOs. Additionally, storing supplementary metadata may enhance processing speeds, albeit at an increased storage requirement.[2]
- *Storage Cost:* The UTXO system generally requires more storage per account due to the need to manage multiple UTXOs. Moreover, every full node is required to retain the complete set of UTXOs to authenticate transactions, which, for instance, translates to each node maintaining a UTXO set size of 3.65 GiB given Bitcoin's current count of approximately 65 million UTXOs. Sophisticated designs like Bulletproofs offer substantial reductions in storage demands by improving the data compactness in cryptographic proofs [4].

4.1.1.1 Types and Properties

Commonly, blockchains are categorized into three distinct types based on their accessibility and scale: public blockchain, private blockchain, and consortium blockchain. A public blockchain is often referred to as a permissionless blockchain. Both private blockchain and consortium blockchain fall under the category of permissioned blockchains.

- *Public Blockchain:* Public blockchains operate as fully decentralized and permissionless networks with self-governing capabilities, such as Bitcoin, Ethereum, and Litecoin. Participation in these networks is unrestricted, allowing anyone to join or exit freely, with no limitations on reading or writing to the ledger. This autonomy ensures the openness and liveness of the network. Nodes within these networks can initiate transactions or engage in the consensus mechanism as miners, validators, or electors. Incentives such as transaction fees and mining rewards are provided to motivate nodes to uphold the network's autonomous operations. Despite these features, public blockchains are vulnerable to various security threats from malevolent entities including, but not limited to, token theft [5], selfish mining [6, 7], and more complex attacks such as the Sybil [8] and eclipse attacks [9]. Consequently, the openness of public blockchains necessitates stringent security measures. Additionally, the visibility of distributed ledgers raises concerns over privacy and anonymity [10], while energy consumption remains a significant issue due to extensive global mining efforts.
- *Private Blockchain:* As indicated by its name, a private blockchain is designed for exclusive use by a particular organization. Access to such a network is

[2] The Bitcoin core version 0.15.0 enhanced block validation performance by allowing a 15% increase in the UTXO database, resulting in a 10% hike in storage needs.

restricted, typically requiring a verified invitation through a secure intranet, making these networks permissioned and closed. Private blockchains boast enhanced efficiency, security, and privacy protection and are easier to manage, though at the cost of reduced decentralization and openness. The central authority in a private blockchain has the capability to alter, rollback, or delete ledger entries, which can compromise the foundational principle of trustlessness in blockchain technology. Private blockchains are particularly advantageous for individual entities seeking to leverage blockchain for specific applications like internal data sharing and identity management.

- *Consortium Blockchain:* Positioned between the fully open public and the restricted private models, consortium blockchains are permissioned and semi-decentralized. Notable examples include Hyperledger Fabric [11], Ripple [12], and Corda [13]. Rather than being controlled by a single entity, consortium blockchains are governed by multiple organizations, which collectively decide on nodes from their respective organizations to participate in the consensus process. Such a structure makes consortium blockchains ideal for facilitating trustless applications across various organizations. For instance, in blockchain-based energy trading scenarios, different stakeholders like energy producers, consumers, and financial institutions can engage in secure and cooperative transactions [14].

Blockchain technology exhibits superior properties that put itself to the front.

- *Immutability:* In essence, immutability implies that once data is recorded on a blockchain, it becomes unchangeable and enduring. This characteristic ensures that modifications to the blockchain, post its initial validation by the network, are impossible. The architecture of blockchain technology, which uses a chained data structure, and the deployment of consensus algorithms that establish agreed-upon, verified states over time, are central to achieving this immutability. This feature is pivotal in preventing disputes and maintaining data integrity, which in turn enhances audit processes. Conversely, a challenge arises in that once private or sensitive information is recorded on the blockchain, it is indelibly public, as historical data is immutable.
- *Decentralization:* Drawing from Paul Baran's seminal 1962 study [15], network structures are primarily categorized into centralized (star) and decentralized (grid or mesh) models. Decentralized networks, resembling a linked set of stars, aptly describe the structure of the Internet. Unlike centralized networks, decentralized systems reduce dependence on any single point of control and are impervious to manipulation by any central authority. Public blockchains, in particular, are engineered to maximize "distributed" characteristics, allowing ledgers to be maintained across a widespread network of nodes. On the other hand, consortium and private blockchains feature a decentralized architecture but with fewer nodes compared to public networks, sometimes bordering on centralization.
- *Openness:* The attribute of openness is inherent to public (or permissionless) blockchains, which are entirely accessible to anyone. Nodes within these networks have the liberty to join or exit freely, conduct transactions, or engage in the

consensus process. This level of openness, unlike traditional banking or service systems, empowers users to possess a copy of the blockchain's historical data, fostering transparency. Moreover, several blockchain projects adopt open-source protocols to enhance public trust and facilitate swift development. However, the absence of a defined threat model, security assurances, and robust implementation could render a blockchain vulnerable and inefficient. Thus, openness not only ensures transparency and auditability but also significantly contributes to the vitality and growth of the blockchain community.

4.1.1.2 Philosophy of Blockchain

In this part, we precisely define the mathematical foundations that support the principles of blockchain technology. The key characteristics of blockchain are derived fundamentally from two critical processes: the creation of honesty and the maintenance of honesty.

Generation of Honesty Consider a group of blockchain nodes represented by d. We introduce an honesty generation function (HGF) that produces a distribution of honesty from d, symbolized as $D = HGF(d)$. Subsequently, by sampling $S(D)$, we find that D adheres to the principle of major honesty with a probability P, such that $P = S(D)$.

The likelihood P is influenced by two main variables: d and S. The composition of d is determined by the registration policies of the blockchain, while S is shaped by the blockchain's operational protocols. A straightforward approach sets S to perform random sampling, making P solely dependent on d, which must meet the criteria for major honesty. Conversely, more complex models might allow for a more flexible d by implementing advanced strategies for S.

Preservation of Honesty Once honesty is established, ensuring its preservation both in time and space is essential. Temporal security underscores the chronological sequencing of actions, accurately documenting events as they transpire. The definitive record following the i-th event e_i is formulated as $R_{i+1} = g(e_1, e_2, \ldots, e_i)$, where $g(\cdot)$ signifies a temporally secure event logging function. Alternatively, R can be expressed recursively: $R_{i+1} = g(R_i, e_i)$. This setup ensures that subsequent events do not improperly precede earlier ones, thereby addressing the atomicity of events. For verification, it is crucial to efficiently confirm the sequential integrity of R.

Nonetheless, a single copy of a blockchain, R, cannot independently assure its security—it could be tampered with if not collectively maintained by multiple nodes. Hence, spatial security is required to safeguard R. The concept of a distributed ledger, inspired by distributed databases, serves this purpose. The storage distribution function, $h(\cdot)$, ensures that $(R^{(1)}, R^{(2)}, \ldots, R^{(n)}) = h(R)$. Here, $R^{(1)}, R^{(2)}, \ldots, R^{(n)}$ represent data held by individual blockchain nodes. Additionally, we define $h^{-1}(\cdot)$ as the function capable of reconstructing R from its parts: $R = h^{-1}(R^{(1)}, R^{(2)}, \ldots, R^{(n)})$. Distributing exact replicas of R across all nodes, i.e., $R^{(1)} = R^{(2)} = \cdots = R^{(n)} = h(R) = R$, although straightforward, results in inefficiency and elevated storage demands. More intricate design of $h(\cdot)$ could reduce storage space while maintaining the integrity of $R^{(1)}, R^{(2)}, \ldots, R^{(n)}$.

4.1.2 Network Requirements for Blockchain

The network layer requirements of blockchain technology can be divided into two parts: one part focusing on consortium and private blockchains, and the other part on public blockchains. First, we will examine the network layer requirements of blockchain protocols for wired networks. Then, we will analyze the network layer requirements of wireless blockchains.

Consortium and Private Blockchains Consortium and private blockchains have network layer requirements primarily focused on security, efficiency, and ease of management. Since these blockchains typically operate in controlled environments with pre-verified and authorized participating nodes, the security demands are lower, allowing the use of dedicated virtual private networks (VPNs) or proprietary networks to ensure data transmission security. The network topology often adopts a centralized or semi-centralized structure, with predetermined connections between nodes, leading to a stable and easily manageable network structure. The bandwidth requirements are low, and latency is minimal, ensuring fast data transmission and processing. Consensus data transmission in the consortium and private blockchains relies on efficient dedicated network channels, reducing the risk of transmission delays and data loss. Additionally, these blockchains can optimize network protocols based on specific application needs, such as using efficient RPC protocols or customized transport layer protocols to enhance data transmission efficiency. In terms of scalability, with a limited and known number of participating nodes, it is easier to achieve horizontal scaling by adding hardware resources or using sharding techniques to improve performance. Reliability is enhanced through multiple backups and fault-tolerant mechanisms.

Public Blockchains Public blockchains (e.g., Bitcoin and Ethereum) have network layer requirements that emphasize decentralization, security, and scalability. Due to the open and unknown nature of participating nodes, highly secure network protocols are needed to prevent man-in-the-middle and distributed denial-of-service (DDoS) attacks. These blockchains use decentralized P2P network topologies, with dynamic and random node connections, and constantly changing network topologies, requiring more complex network discovery and routing protocols. With a large number of widely distributed nodes, network bandwidth and latency become major bottlenecks, necessitating optimized data transmission protocols to reduce delays and bandwidth consumption. Consensus data is transmitted over the public Internet, making it susceptible to network instability and attacks, requiring more robust transmission mechanisms and redundancy designs to ensure data integrity and availability. Public blockchains use standard network protocols (e.g., Transmission Control Protocol/Internet Protocol (TCP/IP)) and optimize them, such as employing advanced encryption techniques and fast node discovery algorithms, to enhance network performance. For scalability, more complex scaling solutions like sharding and Layer 2 solutions are needed to handle the increasing number of nodes and transaction volumes. Reliability relies on the distributed network and redundancy designs, with replication and distributed storage to improve data reliability, necessitating more redundancy and fault tolerance handling at the network layer.

Communication Requirements Effective management of communication resources is a critical component within wireless blockchain networks (WBNs). These resources, encompassing spectrum availability, bandwidth, transmission power, and receiver sensitivity, are essential for ensuring that the network can support high transaction volumes, achieve timely consensus among nodes, and uphold stringent security measures in wireless communications. Efficient management of these resources is crucial for overcoming the inherent limitations of wireless connectivity and meeting the core requirements of scalability, throughput, latency, and security in wireless blockchain networks.

Zhang et al. [16] investigate the communication resource demands required to operate wireless blockchain networks (WBNs) across various consensus mechanisms, including Proof-of-Work (PoW), practical Byzantine fault tolerance (PBFT), and RAFT, specifically in the context of dynamically changing wireless environments. Their research introduces a pioneering analytical framework designed to evaluate the impact of communication resources on key blockchain performance indicators such as security, scalability, throughput, and latency. The authors meticulously analyze the role of communication throughout the stages of blockchain operations—ranging from client requests to consensus, followed by state replication and responses to clients. Their findings reveal that each stage uniquely influences the blockchain's performance, with the consensus process requiring the most significant communication resources due to its necessity for widespread and rapid data dissemination among nodes to achieve synchronization. They demonstrate that the adequacy of communication resources during the state replication phase is critical for maintaining the integrity and timeliness of blockchain updates across the network. Additionally, their analysis highlights that the initial client request and final client response stages, while less demanding, are crucial for initiating and confirming transactions within the network, ensuring the overall reliability and efficiency of blockchain operations. This study stands out by addressing the nuanced communication challenges inherent in blockchain operations and by elucidating the distinct demands and consequences associated with each examined consensus mechanism. Their innovative approach integrates an in-depth communication resource analysis with performance evaluations of WBNs, highlighting the essential role of tailored communication strategies to optimize blockchain efficacy in wireless scenarios.

To substantiate their theoretical framework, Zhang et al. present extensive simulation data alongside analytical results that illustrate the effects of varying communication resource allocations on blockchain network performance. They employed a combination of simulation and analytical modeling to meticulously measure how variables such as spectrum availability, transmission power, and receiver sensitivity influence key performance metrics of blockchain. Their experimental setup was strategically designed to mirror practical wireless blockchain applications, ensuring the relevance and applicability of their findings. Through this robust validation process, the researchers effectively demonstrate the critical link between communication resource management and the efficiency and security of blockchain networks operating in wireless environments.

4.2 Wireless Blockchain Protocols

4.2.1 Blockchain Applications in Wireless Networks

To fully leverage the advantages of blockchain technology, the simplest method is to directly deploy blockchain into wireless networks. Directly deploying blockchain into wireless networks is equivalent to treating the wireless network as if it were a wired network. This approach requires the wireless network to provide the same interface functions as a wired network. It considers the blockchain layer and the network layer as two modular components, where the blockchain layer does not concern itself with the specific implementation of the network layer. This approach enables blockchain technology to provide three significant benefits within the target wireless network domain: access management, security enhancement, and privacy protection. Blockchain ensures more precise and secure allocation and access control of network resources. Its decentralized and immutable nature significantly enhances the overall security of wireless networks, effectively preventing data tampering and malicious attacks. Additionally, blockchain technology offers robust data encryption and privacy protection mechanisms, ensuring the security of users' sensitive information during transmission and storage. These advantages collectively can substantially improve the performance and reliability of wireless networks.

4.2.1.1 Access Management

Access management involves the process of controlling and regulating permissions to access information system resources. It ensures that only verified and authorized individuals can access the system and data in the correct manner and with appropriate restrictions.

Role-based access control (RBAC) is a widely used method that manages user access to system resources by assigning specific permissions to different roles. Multifactor authentication (MFA) adds an extra layer of security by requiring users to provide two or more forms of identity verification before accessing sensitive information.

Blockchain utilizes smart contracts to implement dynamic and programmable access control policies, ensuring the automated enforcement of data access rules. Through a distributed ledger, each access attempt is recorded and verified, enhancing the transparency and traceability of access. Unlike traditional centralized access control systems, blockchain's decentralized nature eliminates the single point of failure risk associated with centralized permission management, thus enhancing the system's tamper resistance and attack resilience.

Novo et al. [17] designed a blockchain-based decentralized access control system for managing Internet of Things (IoT) device permissions. They automated the management and enforcement of access control policies through smart contracts, reducing the reliance on centralized management and improving the system's security and transparency. Their implementation involved a complex system architecture including a management center, blockchain network, and smart contracts, with a Proof-of-Concept deployment. They evaluated the system's ability to handle a large volume of concurrent access requests and demonstrated its scalability and efficiency across multiple management center nodes. This research not only improved the efficiency of IoT device management but also enhanced data security.

Liu et al. [18] proposed an innovative IoT access control model using blockchain and trusted execution environment (TEE) technologies to tackle security challenges such as unauthorized access and overprivileged attacks. Tokoin-Based Access Control (TBAC) introduces "tokoin" (token + coin) to represent access capabilities as cryptographic assets, providing fine-grained and accountable access control defined by the 4W1H policy (who, what, where, when, and how). The system uses blockchain for secure storage, atomic transfer, and logging, while TEE ensures secure execution and monitoring. Operations include creation, transfer, redemption, modification, and revocation of tokoins. TBAC guarantees security, atomicity, auditability, and fine-grained access control, preventing forgery and policy violations. Implemented on Tendermint and Ethereum platforms, TBAC includes a TEE-based control object and demonstrates its practical application in secure, accountable in-home cargo delivery. By materializing access privileges as tokoins, TBAC enhances IoT security, addressing deficiencies in current access control mechanisms.

Xu et al. [19] address the challenges introduced by the complexity of decentralized edge networks. They propose Curb, a novel group-based software-defined networking (SDN) control plane that integrates blockchain and Byzantine fault-tolerant (BFT) consensus to ensure Byzantine fault tolerance, verifiability, traceability, and scalability within a single framework. Curb supports trusted flow rule updates and adaptive controller reassignment, reducing overheads caused by blockchain consensus with a message complexity upper bound by O(N). The architecture employs a group-based approach where each switch is assigned a group of controllers, reducing message complexity and improving scalability. By using a permissioned blockchain running BFT consensus, Curb ensures strong consistency and verifiability of operations. Extensive simulations on the Internet2 network topology validate Curb's design, demonstrating its significant reduction in communication overhead while maintaining high performance and fault tolerance. This innovative integration of blockchain and BFT consensus into a group-based SDN control plane provides a scalable and secure solution for edge computing environments.

4.2.1.2 Security Enhancement

Security enhancement refers to the use of technical means, strategies, and control measures to improve the security of information systems. Its goal is to protect systems from threats such as data breaches, service disruptions, and other security vulnerabilities.

Using firewalls is a common method for enhancing network security. Firewalls monitor and control the flow of data in and out of the network, preventing unauthorized access. Encryption technology is also frequently used to protect data security by encoding data, ensuring that only individuals with the decryption key can access the content.

In blockchain, each transaction must achieve consensus among the majority of nodes in the network before it can be recorded in the ledger. Additionally, transaction data is encrypted during transmission, making data tampering more difficult. Blockchain technology reduces the risk of attacks associated with traditional centralized databases by using a distributed ledger. Without relying on a single data storage point, it becomes harder for attackers to compromise the entire system through a single point of attack.

Malik et al. [20] and colleagues developed a novel blockchain-based trust management system for vehicular ad hoc networks (VANETs) to enhance communication security and reliability. They integrated data sanitization and optimization algorithms (SLE-WOA) to generate keys, combining sea lion optimization and whale optimization algorithms. Additionally, the system employed a dual-layer node trust evaluation process based on rules and machine learning to further ensure the security of transmitted data and the reliability of nodes. In performance validation, Malik et al. demonstrated that their system outperformed existing methods in key sensitivity and rejection rates. For example, their system produced only 17% of original data in key sensitivity analysis, showing significant improvement over traditional methods, which could recover up to 40% of original data.

Guo et al. [21] designed a framework called end-to-end trust network (ETN) by integrating software-defined networking (SDN), network function virtualization (NFV), and blockchain technology to enhance the security of network resource-sharing in IoT environments. They implemented a consortium blockchain to ensure trust among different network entities, using consensus mechanisms and asymmetric encryption algorithms to guarantee data integrity and tamper resistance.

To address the efficiency of on-chain data storage, they designed an on-chain and off-chain collaborative resource management mechanism, enhancing resource processing and response speed while maintaining security. They also employed deep learning algorithms to optimize resource allocation and demand prediction, dynamically adjusting resource allocation in an automated manner to enhance system flexibility and efficiency.

Guo et al. [22] also designed a three-layer architecture system composed of the physical network layer, blockchain edge layer, and blockchain network layer to improve security and authentication efficiency in IoT environments. In the blockchain network layer, they applied an optimized practical Byzantine fault-tolerant (PBFT) consensus algorithm to store authentication data and logs, ensuring trustworthy authentication and data immutability. The blockchain edge layer implemented name resolution and edge authentication services via smart contracts, improving the system's request-handling speed. Furthermore, to enhance communication security, they designed an asymmetric encryption strategy based on elliptic curve cryptography (ECC), effectively preventing attacks on connections between nodes and terminals. Their research also included a caching strategy based on the belief propagation (BP) algorithm, optimizing the data caching process in edge computing nodes, significantly improving data access efficiency and hit rate. In simulation tests, compared with existing edge computing strategies, this caching strategy reduced average latency by 6%–12% and increased hit rate by 8%–14%, demonstrating significant advantages in security and authentication efficiency enhancement.

4.2.1.3 Privacy Protection

Privacy protection involves practices and measures to safeguard personal or group information from unauthorized access, use, or disclosure. This includes implementing appropriate technical and regulatory strategies to ensure the confidentiality, integrity, and availability of personal data.

Data masking is a technique used to protect privacy by altering sensitive data to prevent the exposure of personal information. Differential privacy, on the other hand, adds a certain amount of random noise to published datasets, making it difficult to identify individual data records and thereby protecting the privacy of data subjects.

Blockchain can achieve privacy protection by integrating advanced cryptographic techniques such as zero-knowledge proofs. This allows for the verification of transaction validity without revealing specific transaction details. Compared to conventional privacy protection techniques, blockchain enables user privacy protection without sacrificing data availability. For instance, through anonymization or pseudonymization techniques, blockchain can ensure transaction traceability and non-repudiation while protecting user identity privacy.

Feng et al. [23], in their research, developed an optimized framework for a blockchain-enabled mobile edge computing (MEC) system, enhancing system performance and security through the joint optimization of multiple key aspects.

They first designed an algorithm that jointly optimized user association and data rate allocation, adjusting the connection between each user and base station, as well as data transmission rates iteratively, to balance network load and reduce communication latency. Subsequently, they introduced a binary algorithm for computing resource allocation, effectively distributing the processing capacity of servers to ensure data processing efficiency and timeliness. Additionally, they optimized the selection and scheduling of block producers, improving the efficiency and security of blockchain operations. Simulation results showed that, compared to existing technologies, their scheme could save up to 30% energy in reducing power consumption and decrease delay-to-finality (DTF) time by approximately 25%, significantly enhancing the energy efficiency of the MEC system and the response speed of the blockchain system. These specific performance metrics convincingly demonstrate the superiority and practical value of their approach.

Liu et al. [24] developed an environmental monitoring system based on a trusted execution environment (TEE) to extend blockchain trust from on-chain to off-chain, enhancing privacy protection. They utilized an Arm Cortex-M33 microcontroller to create a continuous monitoring system capable of generating tamper-proof data in real-time. Furthermore, they designed a data consistency protocol to ensure that data is transmitted authentically, continuously, and fault-tolerantly from the TEE system to the blockchain. This solution is particularly suitable for high-security scenarios, such as vaccine tracking, effectively enhancing the trustworthiness and privacy protection of off-chain physical world data. Although the chapter did not provide direct comparative data with other specific schemes, experimental results showed that the system exhibited superior performance in processing time and transmission efficiency.

Guo et al. [25] propose a decentralized privacy-preserving payment scheme named Vehicloak for location-based vehicular services. The chapter introduces the design of Vehicloak, which is the first blockchain-based scheme to preserve location privacy for vehicle payments. It employs a novel cryptographic technique called zk-GSigproof, which integrates zero-knowledge proof with a group signature to ensure privacy and correctness in the payment process. To verify the feasibility of the scheme, the authors implemented it on a private Ethereum blockchain and conducted comprehensive performance evaluations, showing high efficiency in processing payments. The authors also discuss Vehicloak's applicability to other payment scenarios and propose future research directions, including exploring blockchain scalability techniques and developing privacy-preserving solutions for periodic payments.

4.2.2 *Techniques for Improved Performance*

In some situations, due to the scarcity of communication resources in wireless networks, they may not be able to provide the same interface functions as wired networks, such as in wireless sensor networks (WSN). In such cases, even if the wireless network constructs the same interface functions as a wired network, the performance of the upper-layer blockchain would be significantly reduced. Additionally, in other scenarios, wireless networks can construct interface functions that differ from those of wired networks based on their own characteristics. Such interface functions would be more efficient compared to those constructed by treating wireless networks as wired networks. However, this approach requires coordinated design among the blockchain layer, consensus layer, and network layer. Therefore, we will explore how to enhance the performance of wireless blockchain through coordinated design across three layers: the blockchain layer, consensus layer, and network layer.

4.2.2.1 Blockchain Layer

Block access task control (BAC) technology was developed to tackle the escalating security and efficiency challenges in wireless blockchain networks. As blockchain systems expanded in scale, they grappled with issues such as data inconsistencies, network traffic jams, and increased susceptibility to cyberattacks. BAC innovatively resolves these issues by deploying advanced access control protocols that rigorously authenticate node identities and dynamically modulate their access privileges in response to changing network dynamics and node behaviors. This strategic regulation ensures that only verified nodes can engage in the blockchain, thus preserving data integrity and thwarting unauthorized actions like double-spending and network spam.

Li et al. [26] delve into the application of blockchain technology within wireless blockchain networks, particularly emphasizing the consensus process as influenced by the carrier sense multiple access with collision avoidance (CSMA/CA) protocol. Their study confronts the prominent issue of blockchain forking, a condition exacerbated by the randomness inherent in the CSMA/CA backoff counter, which can lead to concurrent block broadcasts and multiple blockchain ledger versions. To address this, the authors introduce a sophisticated scheme known as block access control (BAC), comprising various strategies to effectively manage block mining and transmission. These strategies are specifically designed to minimize redundant forking block generation by pausing mining activities during backoff periods and by discarding superseded forking blocks that occur due to transmission delays. The innovative aspect of their work is the strategic integration of these controls to optimize network performance, enhancing transaction throughput and block utilization.

To empirically substantiate the efficacy of their BAC strategies, Li et al. develop comprehensive Markov chain models that simulate the operational dynamics of wireless blockchain networks under a range of conditions. These models provide a robust mathematical framework for evaluating the impact of each strategy on key performance indicators, such as block discard rates and mining efficiency. Extensive experimental analyses further quantify the performance enhancements brought by the BAC strategies, demonstrating their potential to significantly improve the reliability and efficiency of blockchain operations in wireless environments. The results affirm the strategies' capacity to mitigate blockchain forking issues, thus facilitating the practical integration of blockchain technology across various wireless applications.

4.2.2.2 Consensus Layer

I. Over-the-Air Computation

Over-the-air computation (AirComp) technology is an innovative approach designed to address the inefficiencies in data processing and communication within distributed systems, including blockchain networks. Originally conceptualized to exploit the superposition property of wireless signals for efficient data aggregation, AirComp has found a promising application in enhancing blockchain scalability and reducing latency. In blockchain environments, where the timely and efficient processing of transactions is critical, the typical broadcast and synchronization processes often lead to bottlenecks. AirComp facilitates a paradigm shift by allowing multiple network nodes to simultaneously transmit their data, which are then aggregated in the air and processed collectively. This simultaneous processing significantly cuts down on the time required for consensus and decision-making processes, thus addressing critical speed and scalability issues within blockchain networks.

Xie et al. [27] propose "AirCon," a BFT consensus protocol specifically designed for use in blockchain networks within wireless environments, featuring the novel integration of over-the-air computation (AirComp). Addressing the substantial challenges posed by high resource consumption and limited device capabilities in traditional blockchain applications, the authors ingeniously incorporate lattice coding with AirComp. This strategic integration tackles the issue of efficiently reaching consensus without the excessive communication and computational overhead typical of conventional methods. By facilitating consensus directly at the physical layer during data reception, their approach not only significantly reduces the necessary resources but also optimizes blockchain technology for the unstable and resource-constrained conditions characteristic of wireless networks. This protocol marks a significant departure from traditional blockchain consensus methods, which are resource-intensive and poorly suited to the dynamic and distributed nature of wireless network environments, offering a tailored solution that enhances scalability and efficiency.

To validate the practicality and effectiveness of their protocol, Xie et al. implemented AirCon within an LTE system, addressing key technical challenges such as synchronization, channel estimation, and feedback mechanisms to ensure smooth operation in a wireless infrastructure. Through extensive experimentation and simulations, they showcased the protocol's enhanced performance under a variety of network conditions. Their results demonstrate a marked reduction in communication overhead and computational complexity compared to traditional PBFT protocols, thereby improving the scalability and applicability of blockchain technology in next-generation wireless networks. This thorough evaluation not only confirms the feasibility of AirCon but also underscores its potential to revolutionize blockchain implementations across wireless domains.

II. Enhanced Threshold BLS
In the burgeoning field of blockchain technology, the need for scalable and secure signature schemes has become increasingly crucial. Enhanced threshold Boneh-Lynn-Shacham (T-BLS) technology addresses this imperative by enabling a subset of network participants to generate a valid group signature, without necessitating the involvement of all parties. Originating from the fundamental BLS (Boneh-Lynn-Shacham) signature system, which is renowned for its efficiency and compact signature sizes, T-BLS enhances this framework by introducing threshold capabilities that are essential for decentralized applications. This technology not only reduces the computational overhead associated with traditional signature schemes but also significantly augments security against collusion and attacks. By allowing a group signature to be derived from a subset of signatures, T-BLS facilitates crucial functionalities in blockchain applications, such as consensus mechanisms and transaction validations, thereby solving scalability issues while enhancing trust and integrity within the network.

Zhang et al. [28] developed a Byzantine fault-tolerant blockchain consensus protocol tailored for single-hop wireless networks, addressing the prevalent challenges of network instability and limited resources that often hinder wireless blockchain deployments. Their proposed low-latency and reliable Byzantine fault-tolerant consensus protocol (LRBP) features several innovative elements designed to enhance reliability and efficiency while maintaining low latency. The protocol introduces a randomized credit-based system for block proposer selection to diminish the risk of adversarial influence, incorporates an advanced threshold Boneh-Lynn-Shacham (T-BLS) signature-based voting mechanism to reduce communication overhead, and employs random linear network coding to ensure reliable block transmissions across unstable wireless channels. These strategic innovations allow the protocol to adeptly manage the distinctive challenges posed by single-hop wireless networks, such as frequent transmission errors and dynamic network topology. Designed with robust security measures, LRBP is equipped to withstand various sophisticated security threats, including 51% attacks, sybil attacks, double-spending, and jamming, thus catering to decentralized applications demanding high security and fault tolerance.

To substantiate the efficacy of their protocol, Zhang et al. conducted comprehensive simulations that demonstrate LRBP's performance under diverse network scenarios. They assessed the consensus success probability, the time complexity of achieving consensus, and the protocol's resilience to various attacks. The simulation results affirm that LRBP maintains a high rate of consensus success across environments characterized by significant SNR outage probabilities and prevalent malicious activities. Additionally, their analyses reveal that the protocol sustains low consensus latency and high transaction throughput, efficiently scaling in larger network configurations without notable degradation in performance. These findings highlight LRBP's practical applicability and underscore its potential to significantly enhance the reliability and operational efficiency of blockchain systems in wireless network environments.

III. Signal-to-Interference-plus-Noise Ratio

The signal-to-interference-plus-noise ratio (SINR) model is a mathematical framework used to analyze the quality of a wireless communication link. It measures the signal strength relative to the interference and noise in the communication channel. The SINR model is crucial for understanding the performance of wireless networks, especially in environments with multiple transmitting devices and potential interference. The role of the SINR model in blockchain is as follows:

- **Handling Unreliable Transmissions.** In wireless blockchain systems, reliable communication cannot always be guaranteed due to interference and noise. The SINR model helps in accurately depicting these conditions, allowing the consensus protocol to be designed with a realistic understanding of the network environment.
- **Optimizing Consensus Algorithm.** By using the SINR model, the PoC consensus protocol can efficiently manage communication among devices. It ensures that the leader election and block proposal processes consider the quality of the communication links, reducing the likelihood of transmission failures.
- **Energy Efficiency.** The SINR model allows the protocol to account for the energy consumed during communication. By optimizing the transmission power based on the SINR, the protocol can minimize energy consumption, which is crucial for battery-powered IoT devices in the network.
- **Preventing Forks.** The SINR model helps ensure that only one leader can effectively broadcast a block at a time, as the model accurately predicts which transmissions will be successful. This reduces the chances of multiple leaders proposing blocks simultaneously, thereby preventing forks in the blockchain.
- **Fairness and Robustness.** The model supports a fair leader election process by considering the transmission success probability based on SINR. This ensures that all nodes have an equal opportunity to become the leader, promoting fairness. Additionally, the robustness of the network is enhanced as the protocol can adapt to varying levels of interference and noise.

Zou et al. [29] propose a fast consensus algorithm for permissioned blockchains in wireless networks. The chapter adopts a realistic signal-to-interference-plus-noise ratio (SINR) model to depict unreliable transmissions in wireless channels. Based on this model, a distributed and randomized consensus algorithm is proposed, which can achieve k-times consensus among n devices within $O(k + logn)$ time steps with high probability. The algorithm is asymptotically optimal in terms of time complexity. The authors conduct rigorous theoretical analysis and extensive simulations to validate their method, demonstrating its effectiveness and feasibility in various wireless scenarios.

They address challenges such as interference, energy consumption, and potential forks in wireless blockchain systems by designing a novel Proof-of-Communication (PoC) consensus protocol. The protocol includes four phases: leader election, block proposal, block validation, and chain updation, ensuring efficiency and fairness in the consensus process. Simulation results show that the protocol exhibits linear scalability with the number of consensuses achieved and confirms the time complexity of $O(k + \log n)$, highlighting its efficiency and robustness. The algorithm is particularly useful for large-scale IoT deployments where reliable network support cannot be guaranteed. Its design considers the limited energy resources of wireless devices, making it suitable for battery-powered IoT devices. The randomized leader election and SINR-based communication model improve the robustness and fairness of the consensus process. This work lays the foundation for further research on jamming and Byzantine failures in wireless networks.

4.2.2.3 Network Layer

I. CSMA/CA

In the evolving landscape of blockchain technology, where transaction efficiency and network scalability are paramount, the integration of carrier sense multiple access with collision avoidance (CSMA/CA) offers a compelling solution. Originally developed to manage data transmission in congested wireless network environments, CSMA/CA helps in mitigating collision risks by enabling devices to sense the communication channel for idleness before transmitting data. This becomes particularly significant in blockchain networks that handle a large volume of transactions. By incorporating CSMA/CA, blockchain systems can effectively reduce the likelihood of data collision during transaction broadcasts. This integration enhances the throughput and reliability of blockchain networks, addressing critical bottlenecks in network congestion and ensuring smoother, faster transaction validations. Thus, CSMA/CA not only fortifies wireless communication protocols but also significantly improves blockchain's operational efficiency, paving the way for more robust digital transaction platforms.

Cao et al. [30] investigate the impact of the CSMA/CA protocol on the performance and security of blockchain networks in the context of wireless Industrial Internet of Things (IIoT). Their study focuses on the significant delays in communication introduced by CSMA/CA, which can severely bottleneck blockchain performance. This is particularly critical in terms of transaction confirmation delays, transaction throughput, and security against double-spending attacks. What sets their work apart is the application of a stochastic model to analyze how CSMA/CA affects a directed acyclic graph-based blockchain, specifically the Tangle protocol utilized in IIoT settings. They provide a novel quantification of the protocol's impact on transaction per second (TPS) rates and transaction loss probability under varying network loads. Moreover, their research contributes important insights into the security vulnerabilities, especially concerning the probability of successful double-spending attacks, which is crucial for the effective deployment of blockchain technologies in environments dominated by wireless communications.

To substantiate their theoretical analysis, Cao et al. conduct a series of simulations that demonstrate the real-world implications of CSMA/CA on blockchain performance and security. Through these simulations, they replicate various network conditions to illustrate the protocol's influence on transaction processing and the security framework. The empirical evidence they present not only validates their theoretical models but also highlights the substantial limitations that CSMA/CA imposes on both the efficiency and security of blockchain systems in wireless settings. This rigorous empirical validation offers critical insights, guiding the development of more resilient blockchain infrastructures for wireless IoT applications.

II. Jamming

In the context of wireless blockchain technology, jamming presents a sophisticated challenge, undermining the integrity and performance of the network. Jamming attacks involve the deliberate emission of radio frequency signals to disrupt the regular communication between nodes, thereby compromising the reliability of blockchain transactions. This form of attack is particularly pernicious due to its capacity to induce significant communication delays, increase packet loss, and even partition the network. Such disruptions can severely impact the consensus mechanisms crucial for maintaining blockchain coherence and security. Moreover, jamming can lead to increased energy consumption as nodes boost their transmission power in attempts to counteract interference. Addressing these challenges, researchers and engineers have developed various mitigation strategies, including frequency hopping to evade persistent interference, employing spread spectrum techniques to enhance signal resilience, and implementing sophisticated error correction codes to recover lost data. These defensive measures are essential for preserving the operational efficacy and security of wireless blockchain networks, ensuring that they can withstand and adapt to the complexities of radio-based communication interference.

Xu et al. [31] investigate the security performance of RAFT-based wireless blockchain networks in the presence of malicious jamming, a critical issue in the expansive landscape of wireless IoT ecosystems where interference can jeopardize transaction integrity. Their research focuses on the adaptability of the RAFT consensus mechanism in environments vulnerable to security threats, offering a novel approach to securing blockchain transactions in decentralized networks without relying on traditional, computationally intensive protocols like Proof-of-Work or Proof-of-Stake. The innovative aspect of their work lies in the detailed modeling of blockchain transactions as wireless communications involving both uplink and downlink transmissions. By simulating various jamming scenarios, they develop an analytical model that predicts the probability of successful blockchain transactions, effectively mapping the transaction process onto a communication framework that accounts for signal interference and noise within wireless blockchain setups.

To validate their theoretical models, Xu et al. conducted extensive simulations that replicate real-world interference conditions in wireless communications. They methodically evaluated the impact of varying signal-to-interference-plus-noise ratio (SINR) thresholds on the reliability of blockchain transactions. These simulations not only confirmed the accuracy of their analytical predictions but also demonstrated the practical applicability of their approach in enhancing the reliability of blockchain operations amid adversarial wireless conditions. The detailed experimental validation offers compelling evidence that the RAFT consensus mechanism, when adapted to the challenges of wireless communication, can sustain the integrity of blockchain operations against the dynamics of malicious interferences, marking a significant advancement in the application of consensus algorithms in less controlled, real-world environments.

4.2.3 Examples of Wireless Blockchain Protocols: BLOWN

4.2.3.1 Models and Assumptions

Network Model In this section, we examine a communication network comprising a set V of N nodes distributed randomly within a communication arena. This configuration might include a cluster of remotely controlled unmanned aerial vehicles (UAVs) or smart robots in practical applications. Each node is outfitted with a half-duplex transceiver, capable of either transmitting or receiving messages or monitoring the channel. However, it cannot engage in transmitting and receiving or transmitting and monitoring concurrently. We define $d(u, v)$ as the Euclidean distance between any two nodes u and v. Additionally, $D_R(v)$ represents the circle with radius R centered at node v, and $N_R(v)$ signifies the collection of nodes within the radius R of v, including v itself.

Nodes are presumed to be aware of the identities, locations, and public keys of their counterparts. It is also postulated that each node can create key pairs and utilizes a secure Existential Unforgeability against chosen-message attacks (EUF-CMA) digital signature framework (the cryptographic mechanisms used here are elaborated in sections on protocol analysis and simulation). Nodes maintain a blockchain composed of a sequence of blocks, where each block includes numerous transactions. Key terminology such as transactions, blocks, blockchains, and chains of block headers are denoted as tx, B, BC, and BH, respectively, with additional details indicated through the use of superscripts and subscripts. Transactions are conceptualized as coin exchange mechanisms. We employ the unspent transaction output (UTXO) accounting model. A UTXO is an output from a blockchain transaction that remains unspent, and a UTXO-centric account holds funds across multiple UTXOs. This UTXO framework is characterized by significant scalability and robust security. Owing to the concurrent processing capabilities of UTXOs, they facilitate multi-thread transaction handling, thereby enhancing the scalability of UTXO-based blockchains through techniques like sharding [32]. Moreover, using UTXOs to verify transaction sequences minimizes the risk of double-spending attacks.

Interference Model We utilize the signal-to-interference-plus-noise ratio (SINR) model for wireless networks, which provides a more accurate and realistic representation of network interference compared to models based on graph theory [33].

A standard SINR model can be formulated as follows, which states that a message sent by u is correctly received by v if and only if

$$SINR = \frac{\mathcal{S}}{\mathcal{I} + \mathcal{N}} \geq \beta \quad (4.1)$$

where the received signal power at node v from node u is denoted by $\mathcal{S} = P \cdot d(u, v)^{-\alpha}$, with P representing the consistent transmit power across the network. The interference at node v, $\mathcal{I}$, is given by $\mathcal{I} = \sum_{w \in W \backslash u} P \cdot d(w, v)^{-\alpha}$, where W includes all nodes transmitting during the current round. The ambient noise is represented by $\mathcal{N}$, the path-loss exponent is denoted by α, which lies within the range $(2, 6]$, and the hardware-determined threshold β exceeds 1. For a detailed analysis of noise components, $\mathcal{N} = \mathcal{ADV}(v)$ is defined, where $\mathcal{ADV}(v)$ encompasses noise from environmental and adversarial sources. It is postulated that a uniform noise threshold θ is employed by each node, and the maximum permissible distance between any two nodes is constrained by $R_0 = (P/\beta\theta)^{1/\alpha}$. Moreover, it is assumed that nodes have the capability of performing physical carrier sensing. Should a neighboring node u to v transmit a message, node v will either (a) detect an idle channel if the signal power is below θ; (b) successfully receive a message if the signal power surpasses θ and the $SINR$ is at least β; or (c) detect a busy channel if the signal power is above θ but the $SINER$ is below β. We define $RSS = \mathcal{S} + \mathcal{I} + \mathcal{N}$ as the total received signal power at a node. The combined interference and noise, when a message is received, is calculated as $\mathcal{I} + \mathcal{N} = RSS - \mathcal{S}$, given the known value of $\mathcal{S}$ [34–38].

Epoch-Based Execution As depicted in Fig. 4.2, the BLOWN protocol operates over sequential and nonoverlapping intervals named *epochs*, where a maximum of one block is produced during each epoch. Each epoch, denoted as e_k, is divided into two stages, each comprising several rounds. Within e_k, the notation $r^k 1, i$ refers to the i-th round of the first stage, while $r^k 2, j$ pertains to the j-th round of the second stage. The first stage's rounds include two slots, whereas rounds in the second stage contain a single slot. Additionally, the duration of the first stage is represented by i_k, and that of the second stage is a multiple of the first, $c \cdot i_k$, where c is a defined coefficient to be specified later.

Adversary We postulate the existence of a cohort of adversaries capable of freely entering or exiting the network, forging identities, or creating disturbances to hinder

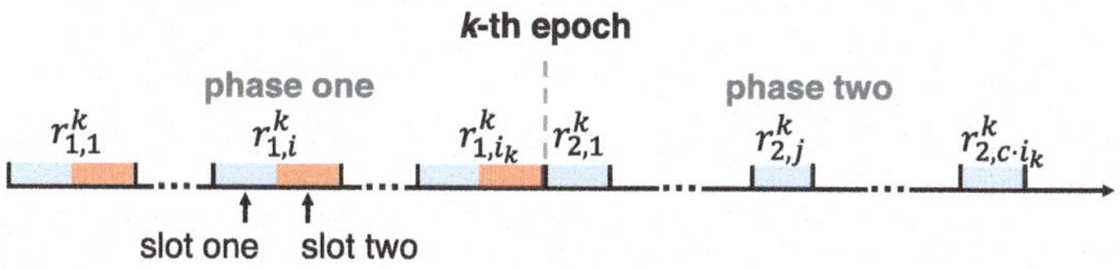

Fig. 4.2 Epoch-based execution

Table 4.1 Summary of notations (BLOWN)

Symbol	Description
B_v^k	Block generated by node v in the k-th epoch
BC_v^k	Blockchain locally stored at node v
BH_v^k	Block headers read from BC_v^k
$P_1(P_2)$	Phase one (phase two)
V	Set of all nodes
N	Network size
c	Constant used to determine the length of phase two
c_v	Counter variable used to record round information
i_k	Length of P_1 of the k-th epoch
tx	A transaction
txp_v	Temporary transaction stack of node v
p_v	v's probability of sending a message
p_V	Aggregated probability of all nodes
$\hat{p}$	An upper bound of p_v
$r_{1,i}^k(r_{2,j}^k)$	i-th (j-th) round of $P_1(P_2)$ in the k-th epoch
l_v	$l_v \in \{0, 1, \cdots, w_v\}$ as the leader counter of v
l_v^0	Initial value of l_v generated by *Sortition*()
T	Time window of the adversary
T_v	Estimate of T by node v
w_v	Deposit of node v
π_v	Proof created by *Sortition*()
ϵ	Proportion of non-jammed rounds
τ	The parameter to determine the hardness of Sortition

any honest nodes at will. For the sake of simplicity, this group is effectively a potent adversary, $\mathcal{A}$, who possesses under 50% of the total network wealth. $\mathcal{A}$ is capable of executing jamming operations. To ensure communication opportunities for honest nodes, $\mathcal{A}$ is constrained as $((1-\epsilon), T)$-bounded throughout any interval spanning T rounds, with $T \in \mathbb{N}$ and $0 < \epsilon \leq 1$. This constraint signifies that the adversary can, at most, disrupt $(1-\epsilon)T$ rounds within any period of T. Moreover, each node v holds a variable T_v, representing its estimate of T.

In this section, we assert that an event E transpires with high probability (w.h.p.) if, for any $c \geq 1$, E is likely to occur with at least $1 - 1/N^c$, and with moderate probability (w.m.p.) if, for any $c \geq 1$, E manifests with a likelihood of at least $1 - 1/\log^c N$. A comprehensive list of all pertinent notations, including those from the BLOWN protocol and its analysis, along with their semantic interpretations, is presented in Table 4.1.

4.2.3.2 The BLOWN Protocol

In this subsubsection, we delineate the dual-phase BLOWN protocol. Initially, we provide a summary of the BLOWN protocol, encompassing a brief on its overall framework and its foundational building blocks, followed by a comprehensive discussion of the protocol specifics.

I. Overview and Utilities of BLOWN

Here, we offer a concise outline of BLOWN, elucidating the fundamental elements and functionalities to more effectively and succinctly delineate the BLOWN protocol.

An Overview on BLOWN The BLOWN protocol operates over periods known as epochs, during which no more than a single block is constructed. Our protocol features two distinct stages, labeled P_1 and P_2 as illustrated in Fig. 4.3. Stage P_1 handles the initialization and selection of a leader, whereas P_2 focuses on the aggregation of transactions and the finalization of blocks. Within our framework, nodes participate by broadcasting on a wireless medium. To counteract interference from both background noise and deliberate jamming, we implement an adaptive transmission strategy, allowing for a jam-resistant communication channel. This adaptation involves modulating the transmission likelihood p_v for each node v, contingent on the perceived network contention. Initially, each node v is assigned a weight l_v through the Sortition algorithm, which takes into account the node's stake, preventing identity proliferation from undermining the protocol. Following this, the leader election phase commences. Leveraging the inherent contention in wireless communications, we design a Proof-of-Channel (PoC) consensus method. This approach enables nodes to vie for channel usage rights, providing them the chance to propose blocks without necessitating additional physical resources or causing extra communication overheads. Specifically, each time a node v receives a message, its weight l_v decreases. The last node with a positive l_v after P_1 concludes is designated

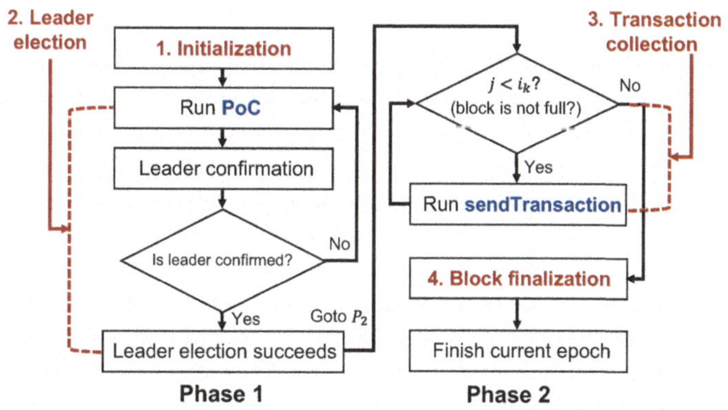

Fig. 4.3 BLOWN protocol overview

the leader, effectively merging the leader selection with channel contention into a unified first phase of BLOWN. In the subsequent phase, P_2, the designated leader gathers and authenticates transactions, compiles them into a block, and broadcasts this block across the network. Upon validation, the block is accepted by all honest nodes.

Utilities Algorithm 1 lists the following frequently used functions for any node v in BLOWN: *Sortition*(), *VerifySortition*(), *LeaderCounter*(), *MSG*(), *MSGT*(), *MSGB*(), *Pack*(), and *Append*(). It also presents the following data structures employed by the above functions: transaction tx, transaction stack txp_v, block B_v^k, block header BH_v^k, blockchain BC_v^k, basic message m, transaction message m_T, and block message m_B.

In a detailed modification, we adapt the cryptographic Sortition algorithm initially developed by Algorand [39] to better align with the needs of our BLOWN protocol. This algorithm utilizes a verifiable random function (VRF) which, upon receiving a private key sk_v, a random seed, and a specified role, generates a hash h_v along with its associated proof π_v. There are two distinct roles involved: a `FOLLOWER`, who participates only as a follower within an epoch, and a `LEADER`, who may potentially lead. Additionally, W represents the combined coin holdings across all users in the network, w_v is the individual deposit of node v, $l_v \in 0, 1, \cdots, w_v$ denotes the leader counter for node v, and $p = \frac{\tau}{W}$ is the probability that each coin contributes to the increment of the counter, where τ is a measure of difficulty. The probability that l_v equals k is calculated following the binomial distribution $B(k; w_v, p) = \binom{w_v}{k} p^k (1-p)^{w_v-k}$, with a total probability of $\sum_0^{w_v} B(k; w_v, p) = 1$. In determining l_v, the *LeaderCounter*(role, w_v, h_v, p) function delineates the unit interval $[0, 1]$ into sequential segments as $I(l_v) = \left[0, \sum_{k=0}^{l_v} B(k; w_v, p)\right]$ for $l_v = 0$ and $I(l_v) = \left(\sum_{k=0}^{l_v-1} B(k; w_v, p), \sum_{k=0}^{l_v} B(k; w_v, p)\right]$ for $I(l_v) \in 1, \cdots, w_v$. If v has the role of `FOLLOWER`, then $I(l_v) = 0$. However, if the normalized hash $\frac{h_v}{2^l}$ (where l is the length of the hash) falls within the interval $I(l_v)$, then l_v is confirmed as the leader counter value. The function *VerifySortition*() is designed to validate whether h_v, π_v, l_v are correct by invoking *VerifyVRF*() and recalculating *LeaderCounter*().

Three functions, namely *MSG*(), *MSGT*(), and *MSGB*(), generate messages that can, respectively, be used for leader election, transaction collection, and block finalization. Specifically, *MSG*() creates a basic message m for leader election in P_1, *MSGT*() produces a message m_T embodying a transaction, which is sent during the transaction collection process in P_2, and *MSGB*() outputs a message m_B which contains a B_v^k generated by the leader v, a BH_v^k read from BC_v^k, the current value l_v of the leader counter, and a string {role, w_v, h_v, π_v, l'_v} used to verify Sortition where l'_v is the original value of the leader counter. To reduce communication cost, we send BH_v^k embodied in m_B for a simplified verification. Finally, *Pack*(txp_v) is adopted to validate and pack transactions to form a new block, and *Append*(BC_v^{k-1}, B_u^k) appends the new block B_u^k to the local blockchain BC_v^{k-1}.

Algorithm 1 Utilities for node v

function *SORTITION*(sk_v, seed, role, τ, w_v, W)
  h_v, π_v = VRF(sk_v, seed||role)
  $p = \tau/W$, l_v =*LeaderCounter*(role, w_v, h_v, p)
  return h_v, π_v, l_v
function *VERIFYSORTITION*(pk_v, h_v, π_v, seed, role, τ, w_v, W, l_v)
  if VerifyVRF(pk_v, h_v, π_v, seed||role) $= 0$ **then**
    return 0
  $p = \tau/W$, $\hat{l_v}$ =*LeaderCounter*(role, w_v, h_v, p)
  if $\hat{l_v} \neq l_v$ **then**
    return 0
  return 1
function *LEADERCOUNTER*(role, w_v, h_v, p)
  $l_v = 0$
  if role is FOLLOWER **then**
    return l_v
  if $\frac{h_v}{2^l} \in \left[0, \sum_{k=0}^{l_v} B(k; w_v, p)\right]$ **then**
    return l_v
  $l_v = 1$
  while $\frac{h_v}{2^l} \notin \left(\sum_{k=0}^{l_v-1} B(k; w_v, p), \sum_{k=0}^{l_v} B(k; w_v, p)\right]$ **do**
    $l_v = l_v + 1$
  return l_v
function $MSG(r_{1,i}^k, l_v)$
  $m.r_{1,i}^k \leftarrow r_{1,i}^k$
  $m.l_v \leftarrow l_v$
  return m
function $MSGT(tx, r_{2,j}^k, l_v)$
  $m_T.tx \leftarrow tx$
  $m_T.r_{2,j}^k \leftarrow r_{2,j}^k$
  $m_T.l_v \leftarrow l_v$
  return m_T
function $MSGB(BC_v^k, B_v^k, r_{2,j}^k, l_v$, role, $w_v, h_v, \pi_v, l_v')$
  $m_B.BH_v^k \leftarrow$ read BH_v^k from BC_v^k
  $m_B.B_v^k \leftarrow B_v^k$
  $m_B.r_{2,j}^k \leftarrow r_{2,j}^k$
  $m_B.l_v \leftarrow l_v$
  $m_B.sort \leftarrow$ {role, w_v, h_v, π_v, l_v'}
  return m_B
function *PACK*(txp_v) ▷ Pack txp to form a block
  return B_v^k
function *APPEND*(BC_v^{k-1}, B_u^k) ▷ append the new block
  return $BC_v^k \leftarrow BC_v^{k-1} + B_u^k$

II. The BLOWN Protocol Specifications

Essentially, BLOWN operates through a dual-phase approach. As detailed in Algorithm 3 for phase P_1 and Algorithm 5 for phase P_2, the first phase, P_1, is dedicated to the setup and selection of a leader, whereas the second phase, P_2, focuses on the gathering of transactions and the completion of block finalization.

Phase P_1 Let us delve into the specifics of the BLOWN P_1 protocol. The steps detailed in lines 2–3 of Algorithm 3 outline the initialization stage. Initially, *Sortition*() is invoked with inputs sk_v, seed||role, τ, w_v, and W (referenced in line 2), producing h_v, π_v, and l_v^0 as outputs. Here, h_v represents a generated hash, π_v is the corresponding proof, and $l_v^0 \in 0, 1, \cdots, w_v$ denotes the preliminary leader counter. It's important to note that $l_v > 0$ signifies that v qualifies as a potential leader, whereas $l_v = 0$ means that v will act as a follower. Let $\hat{p}$ represent the upper limit of transmission probability, set to a minor figure within $(0, 1)$. To avoid the problematic scenario where all nodes might become potential leaders and broadcast simultaneously during the first slot—a situation that happens with a minimal likelihood less than $\hat{p}^n$—we implement a safeguard. This involves ensuring at least one participant is a follower post-initialization. To facilitate this, followers are artificially and randomly incorporated into the network with a designated `FOLLOWBLOWN protocol.` role, effectively managing the network dynamics. Next, we initialize the parameters as follows: $p_v = \hat{p}$, $c_v = 0$, $T_v = 1$, $i = 1$, and $l_v = l_v^0$ (line 3). Here, p_v represents the probability that node v will decide to transmit a message, constrained by the upper limit $\hat{p}$. The variable $c_v = 1$ serves as a counter for recording round information. T_v is the node v's estimate of the adversary's time window, i functions as the round counter for phase P_1, and l_v is the leader counter variable, initialized to l_v^0. Following initialization, phase P_1 progresses through multiple rounds, each comprising two slots, with node v's actions in each slot being contingent upon its designated role.

Before proceeding, it is necessary to clarify the PoC subroutine described in Algorithm 2. This subroutine adjusts the leader counter (l_v), transmission probability (p_v), and the estimate of the adversary's time window (T_v) based on the channel conditions observed during the first slot of each round in P_1. Specifically, a node v with $l_v > 0$ (indicating a potential leader) undertakes the following actions: it either broadcasts a message (m, σ) with probability p_v (lines 1–2), where σ is

---

Algorithm 2 PoC subroutine

```
if v decides to send a message based on p_v then
    m ← MSG(r^k_{1,i}, l_v), v broadcasts (m, σ)
else
    if channel is idle then
        p_v = min{(1 + γ)p_v, p̂}
        T_v = max{1, T_v − 1}
    else
        if v receives a message (m, σ) then
            p_v = (1 + γ)^{-1} p_v
            l_v = l_v − 1
c_v = c_v + 1
if c_v ≥ T_v then
    c_v = 1
    if there is no idle rounds in the past T_v rounds then
        p_v = (1 + γ)^{-1} p_v,
        T_v = T_v + 2
```

Algorithm 3 BLOWN P_1 protocol

1: ▷ Initialization
2: h_v, π_v, l_v^0 =*Sortition*(sk_v, seed||role, τ, w_v, W)
3: $p_v = \hat{p}$, $c_v = 0$, $T_v = 1$, $i = 1$, $l_v = l_v^0$
4: ▷ Leader election
5: **while** TRUE **do**
6: **if** $l_v > 0$ **then** ▷ As a potential leader
7: ▷ slot one of $r_{1,i}^k$
8: run **PoC**
9: ▷ slot two of $r_{1,i}^k$
10: **if** v broadcasts a message in slot one **then**
11: v listens on the channel
12: **if** v senses an idle channel **then**
13: Goto P_2 with $i_k = i$ ▷ run P_2 as a leader
14: **else**
15: $m \leftarrow MSG(r_{1,i}^k, l_v)$, and v broadcasts (m, σ)
16: **else** ▷ As a follower
17: ▷ slot one of $r_{1,i}^k$
18: v listens on the channel to receive a message
19: ▷ slot two of $r_{1,i}^k$
20: **if** in slot one v receives (m, σ) from u and has $\mathcal{I} + \mathcal{N} < \theta$ **then**
21: **if** v senses an idle channel **then**
22: v recognize u as the leader
23: Goto P_2 with $i_k = i$ ▷ run P_2 as a follower
24: **else**
25: $m \leftarrow MSG(r_{1,i}^k, l_v)$, and v broadcasts (m, σ)
26: $i = i + 1$

the signature of m, or it senses the channel with probability $1 - p_v$ (lines 3–10). Node v adjusts p_v in a multiplicative manner, increasing or decreasing by a factor of $(1 + \gamma)$, where $\gamma = O(1/(\log T + \log\log N))$ is a small value influenced by N and T. Specifically, p_v increases multiplicatively (line 5) if the channel is idle and decreases (line 9) when a message is received.[3] This mechanism allows honest nodes to collaboratively adjust their transmission probabilities to mitigate channel contention. Simultaneously, T_v is reduced by 1 if the channel is idle (line 6), suggesting that the adversary's time window estimate might be too high, and l_v is decreased by one if a neighboring node successfully broadcasts a message (line 10), indicating the neighbor has a higher likelihood of becoming the leader. Additionally, if the number of rounds in P_1 is at least T_v (line 12), the protocol checks for any idle rounds in the past T_v rounds (line 14). If none are found, p_v is decreased (line 15) and T_v is increased (line 16) to further refine p_v and T_v. It is evident that a successful broadcast results in the reduction of l_v values among the receivers. When $l_v = 0$,

[3] Receiving a message implies the message has a valid signature; the signature verification process is not explicitly detailed for brevity.

Algorithm 4 sendTransaction subroutine

if v decides to send a message based on p_v **then**
  $m_T \leftarrow MSGT(tx, r^k_{2,j}, l_v)$, and v broadcasts (m_T, σ_T)
else
  if channel is idle **then**
    $p_v = \min\{(1+\gamma)p_v, \hat{p}\}$
    $T_v = \max\{1, T_v - 1\}$
  else
    if receives a message (m_T, σ_T) **then**
      $p_v = (1+\gamma)^{-1}p_v$
$c_v = c_v + 1$
if $c_v \geq T_v$ **then**
  $c_v = 1$
  if there is no idle round in the past T_v rounds **then**
    $p_v = (1+\gamma)^{-1}p_v$,
    $T_v = T_v + 2$

node v transitions to a follower role and only senses the channel in the first slot of the subsequent round.

Now we resume our explanation of Phase P_1 of the BLOWN protocol, which comprises several rounds. During the first slot of each round, if node v is a potential leader (indicated by $l_v > 0$), it executes the PoC subroutine as outlined in Algorithm 2 (line 8); otherwise, it listens to the channel for incoming messages (line 18). In the second slot of each round, node v's behavior is contingent on its actions in the first slot. If v, as a potential leader, broadcasts a message in the first slot and senses an idle channel in the second slot, it can declare itself the leader and proceed to P_2 (lines 10–13); otherwise, it broadcasts a message in the second slot. A follower node v will recognize node u as the leader only if it perceives u as the sole transmitter in the first slot and detects an idle channel in the second slot (lines 20–23); otherwise, it transmits in the second slot (line 25).

At the conclusion of P_1, there should be a single remaining node with $l_v > 0$, which then assumes the role of the leader. Note that i_k represents the duration of P_1, which is used to determine the duration of P_2.

Phase P_2 Phase P_2 of the BLOWN protocol is responsible for transaction collection and block finalization, as outlined in Algorithm 5. This phase comprises a fixed number of $c \cdot i_k$ rounds, with each round containing a single slot. The constant c directly determines the length of P_2 and indirectly influences the maximum block size, which can be adjusted based on specific implementations. In this phase, j denotes the round counter. If $j < c \cdot i_k$, the leader elected in P_1 listens to the channel to receive signed transaction messages, which are then recorded in the stack txp_v, while other nodes continuously broadcast their signed transaction messages (lines 2–8). After completing $c \cdot i_k$ rounds, the leader serializes all transactions to create a new block, denoted by $B^k_v \leftarrow Pack(txp_v)$, and broadcasts the (m_B, σ_B) (lines 12–15). Upon receiving (m_B, σ_B) from node u, node v appends the new block to its local blockchain only if σ_B is valid and *VerifySortition* $(pk_v$, seed, τ, W, $m_B.sort)$

Algorithm 5 BLOWN P_2 protocol

1: ▹ Transaction collection
2: **while** $j < c \cdot i_k$ **do**
3: **if** $l_v > 0$ **then** ▹ As a leader
4: v listens on the channel to receive a (m_T, σ_T)
5: **if** receives m_T.tx$\neq\perp$ **then**
6: $txp_v[j] = m_T.tx$
7: **else** ▹ As a follower
8: run **sendTransaction**
9: $j = j + 1$
10: ▹ Block finalization
11: **if** $j = c \cdot i_k$ **then**
12: **if** $l_v > 0$ **then** ▹ As a leader
13: $B_v^k \leftarrow Pack(txp_v)$
14: $BC_v^k \leftarrow Append(\mathrm{BC}_v^{k-1}, B_v^k)$
15: $m_B \leftarrow MSGB(BC_v^k, B_v^k, r_{2,j}^k, l_v$, role, $w_v, h_v, \pi_v, l_v^0)$, and broadcasts (m_B, σ_B)
16: **else** ▹ As a follower
17: **if** receives (m_B, σ_B) && *VerifySortition* $(pk_v$, seed, $\tau, W, m_B.sort) = 1$ **then**
18: $Append(\mathrm{BC}_v^{k-1}, m_B.B_u^k)$

= 1 (lines 17–18). The *sendTransaction* subroutine detailed in Algorithm 4 is used by P_2 for broadcasting transactions, and the parameters p_v, c_v, and T_v ensure jamming-resistant communication as implemented in the PoC subroutine described in Algorithm 2.

4.2.4 *Examples of Wireless Blockchain Protocols: wChain*

4.2.4.1 Models and Preliminaries

Blockchain Basics Locally, each participant node v possesses a blockchain, represented as BC_v, which is structured as a series of hash-linked blocks. The notation B_v^i symbolizes the i-th block within BC_v. Additionally, we utilize BC_v^{i+} (BC_v^{i-}) to represent the segment of BC_v that precedes (follows) B_v^i. Each block encompasses several transactions; here, tx_i^j designates the j-th transaction in B_v^i. Supposing B_v^k is the most current block on BC_v, the node v's *view* is defined as the tuple $seq, hash_v$, where seq is the sequence number and $hash$ is the hash of B_v^k. Furthermore, the UTXO model is adopted for its notable scalability and robust security features. It is also presupposed that nodes operate under a public key infrastructure, and the cryptographic mechanisms employed, such as digital signatures, are secure, ensuring that no adversary can forge communications.

Network Model The network is conceptualized as a multi-hop wireless ad hoc network comprising a set V of N nodes arrayed over a two-dimensional geographic plane. The term $d(u, v)$ is employed to express the Euclidean distance between any two nodes u and v. We denote $D_R(v)$ as the disk centered at node v with radius R, and $N_R(v)$ represents the collection of nodes, excluding v, that fall within $D_R(v)$. For the sake of simplicity, the minimum distance between any two nodes is standardized to 1, and Γ is the proportion of the maximum distance to the minimum distance between any two nodes. Each node is assigned a unique identifier and is initially aware only of the total number of nodes N. Additionally, the transmission power for each node can be adjusted to reduce interference. Assume that any node may cease to operate at a random point, indicating each node is either active or entirely inactive. A node is classified as *faulty* if it encounters a crash in the current epoch or is missing the most updated view due to failures in previous epochs. Without deviating from generality, it is assumed that N is an odd number. The designed protocol is capable of accommodating up to f faulty nodes, where f is calculated as $f = \lfloor \frac{N}{2} \rfloor$.

Interference and SINR Model The signal-to-interference-plus-noise ratio (SINR) model is utilized to quantify interference within wireless networks, offering a more accurate and realistic depiction compared to graph-based models. The standard expression for the SINR model is given below. According to this model, a message transmitted by node u is successfully received by node v if and only if

$$SINR(u, v) = \frac{P_u \cdot d(u, v)^{-\alpha}}{\mathcal{N} + \sum_{w \in S \setminus \{u\}} P_w \cdot d(w, v)^{-\alpha}} \geq \beta \tag{4.2}$$

holds, where $\alpha \in (2, 6]$ is the path-loss exponent, threshold $\beta > 1$ is determined by hardware, $P_u \cdot d(u, v)^{-\alpha}$ is the received signal power of node v from u, $\mathcal{N}$ is the ambient noise, $P_w \cdot d(w, v)^{-\alpha}$ is the received signal power of node v from w, $S \subseteq V$ denotes the set of nodes transmitting simultaneously with u. Besides, we assume that nodes can perform physical carrier sensing.

Maximal Independent Set (MIS) A subset $S \subseteq V$ qualifies as an *independent set* of V for a given distance r if, for every pair of nodes u and v within S, the distance $d(u, v)$ exceeds r. Moreover, S is deemed a *maximal independent set* if, for any node w not in S, there exists a node x in S for which $d(w, x) \leq r$. The concept of MIS has attracted extensive scholarly interest over recent years, leading to the development of several strategies for computing an MIS through distributed algorithms. In this chapter, we employ the method described in [40], which efficiently calculates a distributed MIS within the optimal duration of $O(\log N)$, provided that the node density remains constant (Fig. 4.4).

Spanner Construction A spanner serves as a core structure for networks, characterized by its minimal linkage and its ability to offer a good balance between efficiency and simplicity, surpassing other network topologies. We utilize these beneficial attributes of a spanner to enhance our data aggregation methodology in *wChain*. Specifically, we implement the distributed spanner construction method outlined in [40]. This algorithm is capable of establishing a sparse spanner, represented as H, which achieves a limited maximum node degree and completes

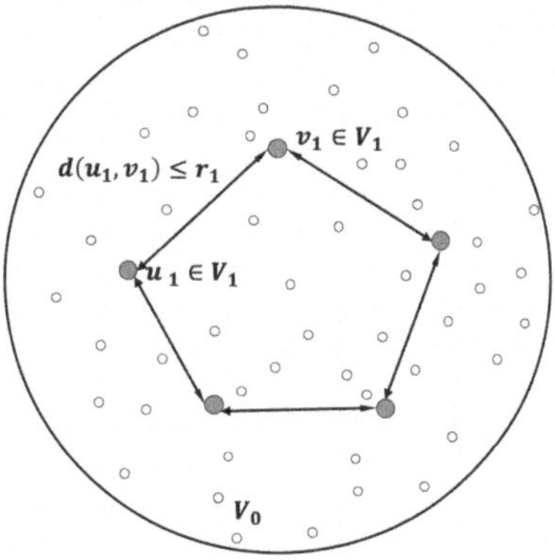

Fig. 4.4 Construction of the first two layers of a spanner followed by an MIS; here hollow and solid circles represent V_0 and V_1, respectively

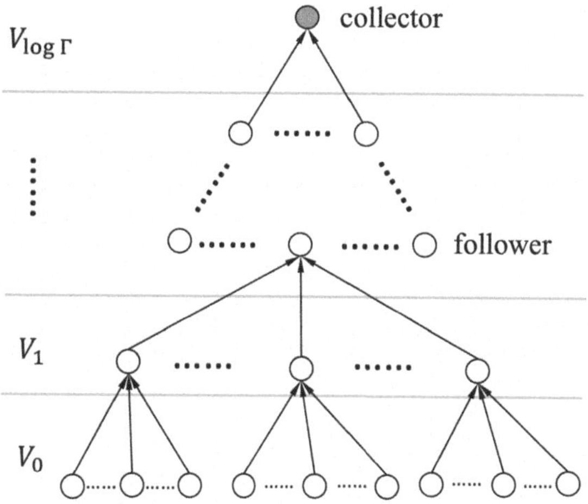

Fig. 4.5 Spanner visualization: A $(\log \Gamma + 1)$-level full spanner where arrows stand for the MIS relationship: each MIS node at a certain level covers a set of nodes in the next lower level

construction within $O(\log N \log \Gamma)$ time slots with high likelihood. As depicted in Fig. 4.5, the developmental stages of H span $\log \Gamma$ rounds; throughout these stages, in the i-th round, where $i = 1, 2, \cdots, \log \Gamma$, the set V_i comprises nodes selected from V_{i-1} forming a maximal independent set. This selection is achieved through the execution of a distributed MIS algorithm based on the distance $r_i = 2^i$. Thus, the sequence progresses as $V_{\log \Gamma} \subseteq \cdots \subseteq V_1 \subseteq V_0 = V$. It is evident that the resultant structure H exhibits the following properties:

- Nodes in V_i constitute an MIS of V_{i-1} with respect to r_i;
- Each node $v \in V_{i-1} \setminus V_i$ has a parent node $u \in V_i$ and $d(v, u) \le r_i$;
- $V_{\log \Gamma}$ contains only one node, i.e., the root.

For clarity in discussion, nodes within $V_0 \setminus V_{\log \Gamma}$ and those in $V_{\log \Gamma}$ are designated as *followers* and *collector*, respectively. In the context of this chapter, we define that an event E transpires with high probability (w.h.p.) if, for any chosen

Table 4.2 Summary of notations ($wChain$)

Symbol	Description
BC_v	The blockchain locally stored at node v
B_v^i	The i-th block in BC_v
BC_v^{i+}	The partial blockchain of BC_v before B_v^i
BC_v^{i-}	The partial blockchain of BC_v after B_v^i
C_v	The set of child nodes of v
d_v	v's data (to be aggregated)
id_v	v's unique id
id_v^p	The unique id of v's parent
M_v	The message queue maintained by v
m_v	A single message sent by v
p	The uniform transmission probability
P_i	The transmission power in the i-th round
$\hat{P}$	The maximum transmission power
R_i	The i-th round
tx_i^j	The j-th transaction in the i-th block
V_i	The set of nodes as an MIS of V_{i-1}
σ	A sufficiently large constant to determine p
λ'	An upper bound of the network density
μ	A constant to determine the round length

$c \geq 1$, the probability of E happening is no less than $1 - 1/N^c$. A detailed enumeration of all essential notations and their meanings is provided in Table 4.2.

4.2.4.2 wChain Protocol

In this segment, we initially outline the design goals and provide a summary of *wChain* along with its associated functionalities; subsequently, we elaborate on the data aggregation and reaggregation procedures, which are integral components of our blockchain framework, and conclude by introducing a robust, three-phase, fast fault-tolerant *wChain* protocol.

I. Design Objectives

The core design principles of *wChain* are aimed at achieving complete decentralization, enhanced efficiency, fault resilience, permanence, and operational continuity.

1. **Full Decentralization:** *wChain* is designed as a distributed system without the need for centralized control, thus eliminating any central point of failure within its architecture.
2. **High Efficiency:** *wChain* is characterized by its low communication overhead, which facilitates rapid processing and enhances transaction throughput.
3. **Fault Tolerance:** *wChain* is engineered to be resilient, capable of handling up to $f = \lfloor \frac{N}{2} \rfloor$ malfunctioning nodes as per the network configuration. It specifically aims to manage the critical issue of node failures during an epoch.

4. **Persistence and Liveness:** Being a blockchain protocol, *wChain* is required to ensure both persistence and liveness, as these terms are described in the section analyzing the protocol.

II. Technical Overview and Utilities

Technical Overview The *wChain* protocol operates in distinct, successive time periods known as *epochs*, with the stipulation that only a single block may be produced during each epoch. Nodes that are operational and not experiencing faults are to add the newly created block to their individual blockchains, thereby collectively preserving a uniform global perspective. At the commencement of each epoch, a spanner is initially set up to serve as the foundational communication network, and the collector of this spanner is designated as the leader for the entirety of the epoch. Should there be a system crash, it necessitates the formation of a new spanner in a reaggregation process within the same epoch. In this situation, the collector may change, however, the leadership remains constant. Data is collected from the followers by the newly appointed collector who then consolidates this data and forwards it to the leader, thereby finalizing the data aggregation for that epoch. The *wChain* protocol advances through three key phases: `PREPARE`, `COMMIT`, and `DECIDE`. During the `PREPARE` stage, the acting leader consolidates view messages from all followers to ascertain the most current consensus. If responses from more than f followers align with the leader's view, the leader proceeds to compile transactions in the `COMMIT` stage. If not, the leader instructs the followers to discontinue the current epoch. The threshold of $(f+1)$ (comprising f followers and one leader) is intentionally set to ensure the *quorum intersection property*, indicating that among these $(f+1)$ nodes, at least one is fault-free and possesses accurate data. In the `DECIDE` phase, the leader assesses the gathered transactions, arranges them into a new block, and disseminates this block along with the view update details to the entire network, enabling the nodes to refresh their local blockchains with the most recent view.

Utilities Before discussing the *wChain* protocol in depth, it is essential to outline the utilities commonly employed as detailed in Algorithm 6. Primarily, the function *MSG*($data_v$) is frequently utilized in *wChain* to create a message m that encapsulates a variable data element, such as a string, a transaction, or a block. For instance, $data_v$ could represent $view_v$ (view information), or tx_v (transaction); it additionally signifies that the provided data is authenticated by v. Besides the primary data content, m also encompasses the `timestamp`, `kindred`, and `role` fields, which are essential for validation purposes. The `kindred` field comprises a tuple that includes the identity id_v of v and the identity id_v^p of v's precursor. The `role` field specifies the role (e.g., follower, collector, or leader) and the exact hierarchy level such as $V_0 \setminus V_1$. A message m is deemed valid by a receiver if it correctly displays both identity and role information. Additionally, M_v denotes a message queue designed to store multiple messages; any repetitive messages within M_v, identifiable via timestamp, are to be removed. A node v can execute an $add(M_v, M_u)$ operation to incorporate M_u into its local M_v. For the process

Algorithm 6 Utilities for node v

function MSG($data_v$)
  $m.data \leftarrow data_v$
  $m.timestamp \leftarrow time$
  $m.kindred \leftarrow \{id_v, id_v^p\}$
  $m.role \leftarrow \{role_v, level_v\}$
  return m
function ADD(M_v, M_u)
  return $M_v \leftarrow M_v \cup M_u$
function PACK(M_v)
  for $m \in M_v$ **do**
    if $m.data$ is a valid transaction **then** organize $m.data$ into B_v
  return B_v
function APPEND(BC_v, B_u)
  if B_u contains the hash of the B_v^k **then**
    return $BC_v \leftarrow BC_v + B_u$
function EXTRACT(BC_v, M_v^{view})
  $i \leftarrow$ the $(f + s)$th highest seq searched in M_v^{view}
  return BC_v^{i+}
function UPDATE(BC_v, BC_u^{i+})
  for $j = 1$ to $|BC_u^{i+}|$ **do**
    if $B_u^j \notin BC_v$ **then** $append(BC_v, B_u^j)$
  return BC_v

of block creation, the operation $pack(M_v)$ is implemented to collate all messages in M_v and systematically compile valid transactions into a block. Furthermore, $append(BC_v, B_u)$ allows for the addition of a block B_u from node u to BC_v only if B_u validates against the hash of BC_v's last block. Moreover, $extract(BC_v, M_v^{view})$ is designed to segregate a partial blockchain BC_v^{i+} from BC_v, where i is defined by the $(f + s)$-th highest seq found in M_v^{view}, with s being a modifiable constant to be specified later. For node revitalization, the function $update(BC_v, BC_u^{i+})$ aids a node v in updating BC_v by integrating missing blocks from the incoming BC_u^{i+}.

III. Data Aggregation and Reaggregation

In this subsection, we discuss two pivotal algorithms essential for executing data aggregation and reaggregation. The main goal of data aggregation is to efficiently gather data from all followers to the leader within a time complexity of $O(\log n \log \Gamma)$ w.h.p. Nevertheless, data loss might occur due to nodes crashing. To address this issue, we introduce a data reaggregation subroutine designed to rectify such disruptions and guarantee that the leader successfully collects complete data from all operational nodes during an epoch.

When implementing blockchain technology in wireless environments, typical unicast and multicast methods are usually unnecessary. It is more advantageous to utilize the inherent broadcast capabilities of the wireless medium. Analyzing the common consensus process reveals three primary communication patterns: one-to-many, many-to-one, and many-to-many. These patterns greatly benefit from the use of broadcast and data aggregation, two fundamental communication primitives.

Broadcasting enhances the efficiency of blockchains, as a one-to-many communication pattern requires only a single broadcast to relay a message from one node to all its peers within the coverage area. In contrast, employing unicast and multicast in many-to-many communications would result in a complexity of $O(N^2)$ for a network with N nodes. However, this complexity can be reduced to $O(N)$ using broadcasting.

Nevertheless, achieving consensus in wireless networks necessitates addressing medium contention and the risk of packet collisions. To mitigate these issues, we set the transmission probability at p and the transmission power at P_i, as outlined in line 6 of Algorithm 7.

Algorithm 7 $DataAggregation(data_v)$ Subroutine

1: **function** DATAAGGREGATION($data_v$)
2:     Initially, $m_v \leftarrow MSG(data_v)$, $M_v = \{m_v\}$
3:     ▹ In $R_i (i = 1, 2, \cdots, \log \Gamma)$:
4:     **if** $v \in V_{i-1} \setminus V_i$ **then**
5:         **for** $\mu \cdot \log N$ slots **do**
6:             send M_v with probability $p = \frac{1}{\sigma\lambda'}$ and power $P_i = 2N\beta r_i^{\alpha}$
7:     **else**
8:         **if** $v \in V_i$ **then**
9:             **for** $\mu \cdot \log N$ slots **do**
10:                 listen on the channel
11:                 **if** receive a valid M_u **then**
12:                     $M_v \leftarrow add(M_v, M_u)$

Data Aggregation Utilizing a spanner, we introduce the *DataAggregation*($data_v$) subroutine to perform hierarchical data consolidation. As depicted in Algorithm 7, a node v implementing *DataAggregation*($data_v$) uses $data_v$ as input, crafts a message m_v via *MSG(data)*, and sets up $M_v = m_v$. The process of data aggregation then unfolds over $\log \Gamma$ rounds. Each round, denoted as R_i, involves nodes v in $V_{i-1} \setminus V_i$ where each has a parent u in V_i satisfying $d(v, u) \leq r_i$. Consequently, round R_i facilitates the transfer of data from $V_{i-1} \setminus V_i$ to V_i, culminating in round $R_{\log \Gamma}$ where the collector receives data from $V_{\log \Gamma - 1} \setminus V_{\log \Gamma}$.

During each specific round R_i, nodes in $V_{i-1} \setminus V_i$ continually transmit M_v for $\mu \cdot \log N$ slots at a probability of $p = \frac{1}{\sigma\lambda'}$, with $\lambda' = 25$ and μ, σ as large constants. This $\mu \cdot \log N$ represents the ideal number of slots to ensure complete data aggregation from all child nodes to their parent node w.h.p., assuming no crash failures. The transmission power is set at $P_i = 2N\beta r_i^{\alpha}$, enabling each node's transmission range to extend to $2^{1/\alpha} r_i$, slightly exceeding r_i. This ensures that for any child node v, its parent is within a unit disk centered at v with radius r_i. This strategy of power control significantly enhances the capability of child nodes to mitigate interference from beyond the unit disk, thereby aiding in successful transmissions. Additionally, a node v in V_i serving as a parent in R_i monitors the channel for $\mu \cdot \log N$ slots to receive messages from its offspring. Should v acquire a M_u from a progeny u, it integrates M_u into M_v. The full cycle of data aggregation, requiring $\mu \log N \log \Gamma$ slots, ensures the collection of data from all functioning nodes.

Algorithm 8 $Reaggregation(data_v)$ Subroutine

▹ as a leader
while $true$ **do**
    ▹ slot one
    broadcast M_ℓ^{data}
    ▹ slot two
    listen on the channel
    ▹ slot three
    if sense noise $> \mathcal{N}$ in slot two **then**
        broadcast $m_\ell \leftarrow MSG(reaggregation_\ell)$
    else
        broadcast $m_\ell \leftarrow MSG(stop_\ell)$ and break
    ▹ data reaggregation
    wait for aggregated data from a collector
▹ as a follower
while $true$ **do**
    ▹ slot one
    listen on the channel
    ▹ slot two
    if receive M_ℓ^{data} in slot one and $data_v \notin M_\ell^{data}$ **then**
        broadcast $m_v \leftarrow MSG(miss_v)$
    ▹ slot three
    listen on the channel
    ▹ data reaggregation
    if receive reaggregation message in slot three **then**
        run $SpannerConstruction$
        run $DataAggregation(data_v)$
    else
        break

Reaggragation The reaggregation subroutine is divided into two phases: a three-slot integrity check phase (lines 3–11, lines 16–22) and a data reaggregation phase (lines 12–13, lines 23–28). We characterize a broadcast operation (implemented in lines 4/9/11/20) as the transmission of a message with $\hat{P} = 2\mathcal{N}\beta r_{\log \Gamma}^{\alpha}$, ensuring that a node monitoring the channel may either receive the message or detect noise surpassing $\mathcal{N}$. Specifically, the integrity check phase aims to verify if the leader has missed any messages from fault-free nodes through physical carrier sensing. In the first slot, the leader l disseminates its current M_ℓ^{data} across the network, where M_ℓ^{data} represents the message queue containing data-type messages (for example, M_ℓ^{view} pertains to view messages). Upon receipt of M_ℓ^{data}, each node v checks whether its $data_v$ is present in M_ℓ^{data}. If absent, v broadcasts $m_v \leftarrow MSG(miss_v)$, enabling the leader in the second slot to identify missing messages by detecting noise levels above $\mathcal{N}$, and in the third slot, the leader broadcasts

$m_\ell \leftarrow MSG(reaggregation_\ell)$ to initiate the second phase. During the data reaggregation phase, all nodes except the leader execute *SpannerConstruction* to rebuild a spanner devoid of faulty nodes. The *SpannerConstruction* method follows the protocol detailed in Sect. 4.2.4.1. The newly formed spanner excludes the leader and selects a new collector responsible for transmitting the compiled data back to the leader. Subsequently, nodes whose messages were omitted in M_ℓ^{data} undertake $DataAggregation(data_v)$. The reaggregation process concludes only when the leader can confirm that no messages from non-faulty nodes are missing, signaled by broadcasting $m_\ell \leftarrow MSG(stop_\ell)$.

Algorithm 9 Fast Fault-Tolerant Blockchain Protocol

▷ PREPARE
▷ as a leader
broadcast $m_\ell \leftarrow MSG(view_\ell)$
listen on the channel for $\mu \log N \log \Gamma$ slots
execute $Reaggregation(view_\ell)$
▷ as a follower
if receive $view_u$ from a leader **then**
  run $DataAggregation(view_v)$
else
  abandon the current epoch
execute $Reaggregation(view_v)$
▷ COMMIT
▷ as a leader
if $|\{m \in M_\ell^{view} | m.data = view_\ell\}| \geq f + 1$ **then**
  broadcast $m_\ell \leftarrow MSG(correct_\ell)$
  listen on the channel for $\mu \log N \log \Gamma$ slots
execute $Reaggregation(tx_v)$
▷ as a follower
if receive $correct_\ell$ from a leader **then**
  run $DataAggregation(tx_v)$
else
  abandon the current epoch
execute $Reaggregation(tx_v)$
▷ DECIDE
▷ as a leader
$B_\ell \leftarrow pack(M_\ell^{tx})$, and $BC_\ell \leftarrow append(BC_\ell, B_\ell)$
broadcast $BC_\ell^{i+} \leftarrow extract(BC_\ell, M_\ell^{view})$
▷ as a follower
if receive BC_ℓ^{i+} from the leader **then**
  update(BC_v, BC_ℓ^{i+})
else
  abandon the current epoch

IV. Fast Fault-Tolerant Blockchain Protocol

wChain is a three-phase protocol developed to establish consensus on the ordering of blocks and to manage disruptions due to failures in wireless network nodes. Specifically, *wChain* utilizes broadcast communications alongside data aggregation and reaggregation techniques, all tailored for the unique challenges of wireless networks. Each epoch in *wChain* advances through three stages: `PREPARE`, `COMMIT`, and `DECIDE`. Below, we elaborate on each stage, illustrating the protocol's capacity for rapid fault tolerance within *wChain*.

Prepare The `PREPARE` phase is designed to assist a leader in establishing a comprehensive overview of the network state. Recall that upon the initial construction of a spanner, *wChain* designates the collector as the leader for overseeing the current epoch. In this capacity, the leader ℓ disseminates $m_\ell \leftarrow MSG(view_\ell)$ across the entire network. Upon receiving this view data from ℓ in the preceding slot, each follower v initiates $DataAggregation(view_v)$. If the view data is not received, the follower will exit the current epoch. It is crucial for all nodes to implement the reaggregation process to guarantee the comprehensive collection of data from all operational nodes.

Commit Define M_ℓ^{view} as the message queue containing the view data of leader ℓ. The criterion that $|m \in M_\ell^{view}|m.data = view_\ell| \geq f + 1$ indicates that the leader must receive at least f view messages corresponding to its own view. Essentially, this requirement confirms that a minimum of $f + 1$ nodes (comprising f followers and one leader) share the same view. Once this condition is met, the leader is authorized to transmit an accurate message and monitor the channel to gather transactions, while followers who discern a valid message commence the process of transaction aggregation. Additionally, the reaggregation subroutine is employed to verify that transactions are fully aggregated.

Decide When the leader ℓ collects all valid transactions from operational nodes, it compiles these transactions into a block B_ℓ and attaches this block to its local blockchain BC_ℓ. Subsequently, the leader ℓ initiates $extract(BC_\ell, M_\ell^{view})$ to construct a BC_ℓ^{i+}, which is instrumental in restoring at least s nodes that experienced failures in earlier epochs and require blockchain updates to regain their non-faulty status. Each operational follower v then updates its own blockchain BC_v using the procedure $update(BC_v, BC_\ell^{i+})$.

4.3 Summary

This chapter provides a detailed description of wireless blockchain protocols. We introduce the basic concepts of blockchain. Blockchain protocols were initially developed and deployed in wired networks and later introduced to wireless networks. Therefore, at the end of the blockchain introduction, we discuss the underlying network requirements for blockchain. If wired blockchain protocols are to be directly transplanted into wireless networks, at a minimum, the wireless network must meet the basic standards required for deploying blockchain in wired networks. Currently, there are two approaches in the industry and academia for designing and applying wireless blockchain protocols. One approach is to directly deploy blockchain protocols into wireless networks to leverage blockchain features to add new characteristics or accomplish specific tasks in application scenarios, such as access management, security enhancement, and privacy protection. The other approach aims to improve the performance of wireless blockchain protocols. This type of research hinges on the differences between wireless and wired networks. Therefore, we introduce technologies and work aimed at optimizing wireless blockchain performance across three layers: blockchain layer, consensus layer, and network layer. Finally, we present two wireless blockchain protocols that perform well in wireless networks: BLOWN and wChain.

References

1. Katz, J., & Lindell, Y. (2014). *Introduction to modern cryptography*. CRC Press.
2. Merkle, R. C. (1987). A digital signature based on a conventional encryption function. In *Conference on the Theory and Application of Cryptographic Techniques* (pp. 369–378). Springer.
3. Becker, G. (2008). Merkle signature schemes, merkle trees and their cryptanalysis. Ruhr-University Bochum, Tech. Rep.
4. Bünz, B., Bootle, J., Boneh, D., Poelstra, A., Wuille, P., & Maxwell, G. (2018). Bulletproofs: Short proofs for confidential transactions and more. In *2018 IEEE Symposium on Security and Privacy (SP)* (pp. 315–334). IEEE.
5. Atzei, N., Bartoletti, M., & Cimoli, T. (2017). A survey of attacks on ethereum smart contracts (sok). In *International Conference on Principles of Security and Trust* (pp. 164–186). Springer.
6. Eyal, I., & Sirer, E. G. (2018). Majority is not enough: Bitcoin mining is vulnerable. *Communications of the ACM, 61*(7), 95–102.
7. Sapirshtein, A., Sompolinsky, Y., & Zohar, A. (2016). Optimal selfish mining strategies in bitcoin. In *International Conference on Financial Cryptography and Data Security* (pp. 515–532). Springer.
8. Douceur, J. R. (2002). The sybil attack. In *International Workshop on Peer-to-Peer Systems* (pp. 251–260). Springer.
9. Singh, A., et al. (2006). Eclipse attacks on overlay networks: Threats and defenses. In *Proceedings IEEE INFOCOM*. Citeseer.
10. Li, X., Jiang, P., Chen, T., Luo, X., & Wen, Q. (2017). A survey on the security of blockchain systems. *Future Generation Computer Systems, 107*, 841–853.
11. Androulaki, E., Barger, A., Bortnikov, V., Cachin, C., Christidis, K., De Caro, A., Enyeart, D., Ferris, C., Laventman, G., Manevich, Y., et al. (2018). Hyperledger fabric: a distributed operating system for permissioned blockchains. In *Proceedings of the Thirteenth EuroSys Conference* (p. 30). ACM.
12. Schwartz, D., Youngs, N., Britto, A., et al. (2014). The ripple protocol consensus algorithm. *Ripple Labs Inc White Paper*, 5.
13. Brown, R. G., Carlyle, J., Grigg, I., & Hearn, M. (2016). Corda: an introduction. *R3 CEV, August*.
14. Li, Z., Kang, J., Yu, R., Ye, D., Deng, Q., & Zhang, Y. (2017). Consortium blockchain for secure energy trading in industrial internet of things. *IEEE Transactions on Industrial Informatics, 14*(8), 3690–3700.
15. Baran, P. (1964). On distributed communications networks. *IEEE Transactions on Communications Systems, 12*(1), 1–9.
16. Zhang, L., Xu, H., Onireti, O., Imran, M. A., & Cao, B. (2021). How much communication resource is needed to run a wireless blockchain network? *IEEE Network, 36*(1), 128–135.

17. Novo, O. (2018). Blockchain meets iot: An architecture for scalable access management in iot. *IEEE Internet of Things Journal, 5*(2), 1184–1195.
18. Liu, C., Xu, M., Guo, H., Cheng, X., Xiao, Y., Yu, D., Gong, B., Yerukhimovich, A., Wang, S., & Lyu, W. (2023). Tbac: A tokoin-based accountable access control scheme for the internet of things. *IEEE Transactions on Mobile Computing, 23*(5), 6133–6148.
19. Xu, M., Wang, C., Zou, Y., Yu, D., Cheng, X., & Lyu, W. (2022). Curb: Trusted and scalable software-defined network control plane for edge computing. In *2022 IEEE 42nd International Conference on Distributed Computing Systems (ICDCS)* (pp. 492–502). IEEE.
20. Malik, N., Nanda, P., He, X., & Liu, R. P. (2020). Vehicular networks with security and trust management solutions: proposed secured message exchange via blockchain technology. *Wireless Networks, 26*, 4207–4226.
21. Guo, S., Hu, X., Guo, S., Qiu, X., & Qi, F. (2019). Blockchain meets edge computing: A distributed and trusted authentication system. *IEEE Transactions on Industrial Informatics, 16*(3), 1972–1983.
22. Guo, S., Qi, Y., Yu, P., Xu, S., & Qi, F. (2020). When network operation meets blockchain: An artificial-intelligence-driven customization service for trusted virtual resources of iot. *IEEE Network, 34*(5), 46–53.
23. Feng, J., Yu, F. R., Pei, Q., Du, J., & Zhu, L. (2020). Joint optimization of radio and computational resources allocation in blockchain-enabled mobile edge computing systems. *IEEE Transactions on Wireless Communications, 19*(6), 4321–4334.
24. Liu, C., Guo, H., Xu, M., Wang, S., Yu, D., Yu, J., & Cheng, X. (2022). Extending on-chain trust to off-chain–trustworthy blockchain data collection using trusted execution environment (tee). *IEEE Transactions on Computers, 71*(12), 3268–3280.
25. Guo, Y., Wan, Z., Cui, H., Cheng, X., & Dressler, F. (2022). Vehicloak: A blockchain-enabled privacy-preserving payment scheme for location-based vehicular services. *IEEE Transactions on Mobile Computing, 22*(44), 6830–6842.
26. Li, Y., Cao, B., Liang, L., Mao, D., & Zhang, L. (2021). Block access control in wireless blockchain network: Design, modeling and analysis. *IEEE Transactions on Vehicular Technology, 70*(9), 9258–9272.
27. Xie, X., Hua, C., Hong, J., Gu, P., & Xu, W. (2023). Aircon: Over-the-air consensus for wireless blockchain networks. *IEEE Transactions on Mobile Computing, 23*(5), 4566–4582.
28. Zhang, L., Zhang, B., & Li, C. (2023). An efficient and reliable byzantine fault tolerant blockchain consensus protocol for single-hop wireless networks. *IEEE Transactions on Wireless Communications, 23*(3), 1974–1987.
29. Zou, Y., Xu, M., Yu, J., Zhao, F., & Cheng, X. (2021). A fast consensus for permissioned wireless blockchains. *IEEE Internet of Things Journal, 10*(14), 12102–12111.
30. Cao, B., Li, M., Zhang, L., Li, Y., & Peng, M. (2019). How does csma/ca affect the performance and security in wireless blockchain networks. *IEEE Transactions on Industrial Informatics, 16*(6), 4270–4280.
31. Xu, H., Zhang, L., Liu, Y., & Cao, B. (2020). Raft based wireless blockchain networks in the presence of malicious jamming. *IEEE Wireless Communications Letters, 9*(6), 817–821.
32. Ozyilmaz, K. R., Patel, H., & Malik, A. (2018). Split-scale: Scaling bitcoin by partitioning the UTXO space. *CoRR*, abs/1809.08473.
33. Yu, D., Zou, Y., Xu, M., Xu, Y., Zhang, Y., Gong, B., & Xing, X. (2020). Competitive age of information in dynamic iot networks. *IEEE Internet of Things Journal, 8*(20), 15160–15169.
34. Ikuno, J. C., Pendl, S., Šimko, M., & Rupp, M. (2012). Accurate sinr estimation model for system level simulation of lte networks. In *2012 IEEE International Conference on Communications (ICC)*, (pp. 1471–1475). IEEE.
35. Zhao, Y., Wu, J., & Lu, S. (2011). Efficient sinr estimating with accuracy control in large scale cognitive radio networks. In *2011 IEEE 17th International Conference on Parallel and Distributed Systems* (pp. 549–556). IEEE.
36. Olszewski, K. J. (2007). Sinr measurement method for ofdm communications systems, August 21, 2007. US Patent 7,260,054.

37. Jeske, D. R., & Sampath, A. (2007). Method of estimating a signal-to-interference+ noise ratio (sinr), February 27, 2007. US Patent 7,184,497.
38. Abu-Dayya, A. (1999). Determining sinr in a communications system, November 23, 1999. US Patent 5,991,273.
39. Gilad, Y., Hemo, R., Micali, S., Vlachos, G., & Zeldovich, N. (2017). Algorand: Scaling byzantine agreements for cryptocurrencies. In *Proceedings of the 26th Symposium on Operating Systems Principles* (pp. 51–68).
40. Yu, D., Ning, L., Zou, Y., Yu, J., Cheng, X., & Lau, F. C. M. (2017). Distributed spanner construction with physical interference: Constant stretch and linear sparseness. *IEEE/ACM Transactions on Networking, 25*(4), 2138–2151.

Chapter 5
Applications of Wireless Consensus

No single wireless consensus algorithm can meet the needs of all application scenarios. In different application contexts, selecting the appropriate wireless consensus algorithm can significantly enhance the efficiency of the entire system, whether in production environments or daily life. The correct choice not only improves efficiency but also enhances the security and reliability of the system. Conversely, choosing an inappropriate wireless consensus algorithm can introduce serious security risks. For example, using a fault-tolerant wireless consensus algorithm in a system with Byzantine nodes may prevent the system from defending against malicious node attacks, leading to data breaches, system crashes, and other security issues. Therefore, it is crucial to accurately select and deploy the corresponding wireless consensus algorithm based on the specific requirements of the application scenario. This ensures that the technological advantages are fully utilized and the system operates efficiently and securely.

This chapter will introduce the application scenarios of fault-tolerant wireless consensus, Byzantine fault-tolerant wireless consensus, and wireless blockchain, respectively. Before discussing the application scenarios, we will first introduce their communication needs and computation requirements. For each application scenario, we will begin by describing their network conditions and consensus requirements, followed by a discussion of some relevant work in the application scenario.

5.1 Overview

The essence of wireless consensus lies in ensuring data consistency and system reliability within wireless networks, particularly when facing network instability and node failures. It encompasses several key aspects:

- **Fault-Tolerant Wireless Consensus:** Ensures that the system can continue to operate and maintain data integrity even if some nodes fail due to technical issues.

M. Xu et al., *Wireless Consensus*, Wireless Networks,
https://doi.org/10.1007/978-3-031-70859-6_5

- **Byzantine Fault-Tolerant Wireless Consensus:** Operates in more stringent environments, maintaining normal network operations and preventing the spread of erroneous information and data tampering, even in the presence of malicious nodes.
- **Wireless Blockchain Protocol:** Enhances data transparency and security through decentralized data management, particularly suitable for applications requiring high trust and security.

In wireless communication, signal interference, high node mobility, and complex, variable network environments often lead to unstable data transmission and node failures. Wireless consensus ensures through algorithms that all nodes in the network can reach a consistent state even in such unstable environments, thus guaranteeing the continuous operation of the entire system and the accuracy of the data. Additionally, this technology reduces dependency on centralized control through distributed decision-making, enhancing the network's resistance to interference while also improving system scalability and self-recovery capabilities. The importance of wireless consensus can be introduced in three main aspects:

Broad Applications Wireless consensus technology has permeated many important fields, improving operational efficiency and security across various industries, proving its wide applicability and criticality. In intelligent transportation systems, wireless consensus technology ensures real-time data synchronization between autonomous vehicles, allowing them to coordinate safely and reduce traffic accidents. In smart grid applications, wireless consensus helps achieve efficient energy distribution, optimizing power production and consumption through real-time data processing and reducing energy waste. In the healthcare sector, it demonstrates its ability to protect the integrity and privacy of sensitive data, ensuring the secure transmission of patient information over networks.

Enhancing Efficiency Wireless consensus technology not only addresses existing problems but also promotes the innovation and development of new technologies. In the Internet of Things (IoT) sector, wireless consensus mechanisms support large-scale device autonomous coordination and data processing through distributed network management, significantly enhancing the efficiency and flexibility of the entire system. This technology allows numerous sensors and devices to communicate effectively and manage themselves without a central server. For example, in smart home systems, various devices such as lights, thermostats, and security systems can exchange data in real time through wireless consensus mechanisms, achieving intelligent environmental adjustments and security monitoring. The application of this technology greatly improves the response speed of operations, reduces system maintenance costs, and enhances user experience. Additionally, the wireless blockchain protocol increases data transparency and system trust by achieving decentralized data storage and processing within the network. This not only improves the data processing and coordination capabilities of large-scale device networks but also provides users with higher operational response speeds and system reliability.

Solving Traditional Challenges Wireless consensus technology offers a novel perspective for addressing many challenges present in traditional wireless networks. Through algorithm and protocol innovations, it effectively resolves issues related to node synchronization, data consistency, and network security. For instance, by implementing wireless consensus algorithms, the network can ensure that information remains accurate and consistent even during network congestion or malicious attacks, thus safeguarding normal network operations and data security. Moreover, the application of this technology helps improve network scalability and flexibility, allowing the network to adapt to changing environments and demands, thereby enhancing overall network performance and reliability.

5.1.1 Requirements for Fault-Tolerant Wireless Consensus

Fault-tolerant wireless consensus is primarily applied in network environments where nodes may fail due to technical issues or outages. Its goal is to ensure that the system can continue to operate stably even when some nodes fail. Below is a detailed discussion of its specific requirements in terms of communication needs and computation requirements.

I. Communication Requirements
The communication needs of fault-tolerant wireless consensus focus mainly on network stability and data transmission between nodes. Since the goal of this technology is to maintain normal system operations even when some nodes fail, the underlying network must support the following characteristics:

- **High Reliability:** The underlying network must have high reliability to minimize communication interruptions between nodes. In fault-tolerant wireless consensus, each node needs to maintain continuous communication with other nodes to ensure data consistency and overall system stability. Therefore, the network must be capable of handling high-load communications and quickly recovering and reestablishing connections in the event of single-point failures.
- **Low Latency:** Data transmission requires low latency to ensure that other nodes can quickly detect and respond when a node fails. Low-latency communication is particularly important for fault-tolerant wireless consensus because the system needs to detect and respond to node failures as quickly as possible to prevent data inconsistencies and system crashes.
- **Redundant Paths:** To cope with node outages, the network must have redundant communication paths to ensure that data can be transmitted through alternative paths if one path fails. This redundancy mechanism not only increases system reliability but also ensures that the system can continue to operate normally and maintain data transmission continuity and stability even when some nodes or communication paths fail.

II. Computation Requirements

In terms of computational power, fault-tolerant wireless consensus typically does not involve complex cryptographic computations. Its main computation requirements are concentrated in the following aspects:

- **State Updates:** Calculating node states and updating consensus states involves relatively simple computations. Each node needs to quickly process and update its state upon receiving new state information to ensure consistency with other nodes. These computations usually do not involve complex algorithms but require fast and accurate processing of large amounts of real-time data.
- **Error Detection:** Simple error detection algorithms are used to identify and handle node outages. The system needs to continuously monitor the operational status of each node and quickly take action (such as reallocating tasks or activating backup nodes) when a node failure is detected to maintain normal system operations. These error detection algorithms are usually based on heartbeat mechanisms or periodic status reports and have a light computational burden.
- **Data Synchronization:** The computational requirements for data synchronization between nodes is low, primarily ensuring data consistency and integrity. During data synchronization, each node needs to periodically exchange data with other nodes to ensure that all nodes' data are up-to-date and synchronized. These synchronization operations typically involve simple data replication and consistency checks, not requiring complex computational resources.

5.1.2 Requirements for Byzantine Fault-Tolerant Wireless Consensus

Byzantine fault-tolerant wireless consensus is primarily applied in complex environments where malicious node attacks are prevalent. Its goal is to ensure that the system can maintain normal operation and data consistency even in the face of frequent malicious node attacks. Below is a detailed discussion of its specific requirements in terms of communication and computational power needs.

I. Communication Requirements

Byzantine fault-tolerant wireless consensus has higher communication requirements, especially when facing malicious node attacks. The underlying network needs to have the following characteristics:

- **High Bandwidth:** Byzantine fault-tolerant consensus typically requires higher bandwidth due to its greater message complexity compared to fault-tolerant consensus. Since a large number of messages need to be transmitted to verify the authenticity of nodes and the reliability of messages, the network bandwidth requirement is high. Byzantine fault-tolerant consensus mechanisms usually involve complex multi-round message exchanges to ensure that all nodes reach consensus. This means the underlying network must have sufficient bandwidth to support high-frequency, high-volume data communication.

- **Wide Coverage:** Ensuring that all nodes can communicate with each other to prevent Byzantine nodes from isolating honest nodes. Byzantine fault-tolerant consensus mechanisms require all nodes to participate in the consensus process, so the network coverage must be wide enough to ensure that each node can reliably communicate with other nodes, preventing consensus failure due to node isolation.
- **Redundant Paths:** To cope with node failures, the network must have redundant communication paths to ensure that data can be transmitted through alternative paths if one path fails. This redundancy mechanism not only increases system reliability but also ensures that the system can continue to operate normally and maintain data transmission continuity and stability even when some nodes or communication paths fail. Given the potential for adversarial attacks, redundant paths are particularly important to ensure that data can still be transmitted through other paths if one path is attacked or fails.

II. Computation Requirements

Byzantine fault-tolerant wireless consensus has high computation requirements, primarily involving complex cryptographic tools:

- **Digital Signatures:** Used to verify the source and authenticity of messages to prevent message tampering. Each node needs to digitally sign messages when sending them, and the receiving nodes need to verify the authenticity of these signatures. This process requires significant computational resources to generate and verify digital signatures, ensuring the reliability of each message.
- **Encryption Algorithms:** Ensure the confidentiality of communication to prevent sensitive data from being intercepted by malicious nodes. In Byzantine fault-tolerant consensus mechanisms, all transmitted data must be encrypted to prevent malicious nodes from intercepting and tampering with it. This requires each node to have efficient encryption and decryption capabilities to handle a large volume of encrypted communication in a short time.
- **Advanced Cryptographic Tools:** Some consensus protocols, especially in asynchronous consensus, commonly use advanced cryptographic tools such as threshold encryption, threshold signatures, and erasure codes. These tools ensure that the system can maintain data security and consistency even in the face of complex attacks. Threshold encryption and threshold signatures allow multiple nodes to participate in the encryption and signing process, enhancing system security, while erasure codes are used to increase data redundancy and recovery capability, ensuring that even if part of the data is lost, the complete information can still be recovered.

5.1.3 Requirements for Wireless Blockchain

Wireless blockchain protocols combine the advantages of wireless networks and blockchain technology, primarily applied in scenarios that require the transparency and public nature of blockchain ledgers. These scenarios typically need historical information to operate and verify data. Below is a detailed discussion of its specific requirements in terms of communication and computational power needs.

I. Communication Requirements
Wireless blockchain protocols, in addition to basic network stability, need to support the characteristics of decentralized data management:

- **Higher Bandwidth:** Compared to the other two consensus protocols, transmitting blocks require higher bandwidth. Blockchain networks require each node to quickly receive and verify newly generated blocks, which imposes high demands on the network's transmission capacity.
- **Low Latency:** Although blockchain systems can tolerate some latency, the network should have low latency to ensure fast verification and confirmation of transactions and blocks. Low latency helps improve the overall efficiency and user experience of the blockchain network, ensuring that transactions can quickly reach consensus and be recorded on the chain.
- **High Reliability:** Data in a blockchain network needs highly reliable transmission to ensure that each node receives complete and consistent block information. Any data loss or error could lead to blockchain inconsistencies, so the underlying network must provide stable and reliable data transmission services.

II. Computation Requirements
The computation requirements of wireless blockchain protocols focus mainly on cryptography and data processing:

- **Hash Computation:** Hash functions are extensively used to ensure data integrity and immutability. Each newly generated block requires hash computation to ensure the integrity and security of the block's contents. This requires each node to have efficient hash computation capabilities to quickly process and verify block data.
- **Digital Signatures and Encryption:** These are used to protect the security of transaction data, preventing forgery and tampering. In a blockchain system, every transaction needs digital signing and encryption to ensure its legality and security. Each node must have efficient signing and encryption processing capabilities to verify a large number of transactions in a short time.
- **Consensus Algorithms:** Running consensus algorithms such as Proof-of-Work (PoW) and Proof-of-Stake (PoS) typically involves intensive computation, especially PoW, which requires substantial computational power to solve hash puzzles. Blockchain consensus mechanisms require each node to participate in complex computational processes to ensure the system's decentralization and security. This places high demands on the nodes' computational capabilities, particularly under the PoW mechanism, where significant computational power is needed for hash collision calculations. If fault-tolerant wireless consensus or Byzantine fault-tolerant wireless consensus is used in the wireless blockchain protocol, corresponding computational power is also needed to support these mechanisms.

5.2 Application Scenarios of Wireless Consensus

The application scenarios of wireless consensus technology in modern communication networks are broad and diverse, covering multiple key areas. In fault-tolerant wireless consensus applications, the system can continue to operate when nodes experience failures such as downtime, ensuring the continuity and reliability of communication. In Byzantine fault-tolerant wireless consensus applications, the system can maintain correct operation despite frequent malicious attacks, preventing data tampering and malicious behavior from affecting network stability. Wireless blockchain application scenarios leverage the transparency and openness of blockchain to ensure traceability of historical information, suitable for applications requiring high trust and transparency, such as supply chain management, smart cities, and the Internet of Things. Additionally, these application scenarios encompass areas such as edge computing and mobile payments, demonstrating the broad application potential of wireless consensus technology in different environments.

5.2.1 Intelligent Transportation Systems

In intelligent transportation systems (ITS), the application of fault-tolerant wireless consensus, Byzantine fault-tolerant wireless consensus, and wireless blockchain protocols can significantly enhance the reliability, security, and data integrity of the system. Fault-tolerant wireless consensus ensures that the nodes in the transportation network can maintain consistent decision-making in the face of communication failures or node malfunctions, thereby preventing traffic chaos and improving traffic flow smoothness and safety. Byzantine fault-tolerant wireless consensus further strengthens the security of ITS, particularly in the presence of malicious attacks or nodes intentionally sending erroneous information. Through complex verification mechanisms, it ensures that other nodes can still reach a consensus decision, preventing potential traffic accidents. Wireless blockchain protocols are widely used in ITS, especially in the secure transmission and storage of data. The decentralized and tamper-proof characteristics of blockchain technology ensure the security and trustworthiness of data exchange between vehicles and between vehicles and infrastructure. For instance, vehicle driving data and traffic accident records can be recorded on the chain in real time, preventing data tampering and providing transparent and reliable traffic information management.

I. Network Conditions

- **Common Devices:** Key devices in intelligent transportation systems include autonomous vehicles, roadside units (RSUs), traffic signal controllers, and various sensors. Autonomous vehicles use onboard sensors and computing units for environmental perception and decision-making, while RSUs and traffic signal controllers manage traffic flow and signal lights.

- **Wireless Communication Protocols:** Common wireless communication protocols include dedicated short-range communications (DSRC), cellular vehicle-to-everything (C-V2X), Wi-Fi, and 5G. DSRC and C-V2X are specifically designed for vehicular networks, featuring low latency and high reliability, making them suitable for communication between vehicles and between vehicles and infrastructure.
- **Network Architecture:** Intelligent transportation systems typically use multi-hop network architecture with dynamic topology. In scenarios with base stations (like C-V2X) and without base stations (like DSRC), vehicles can relay data through multiple hops to ensure communication stability and coverage.

II. Consensus Requirements

- **Latency:** Intelligent transportation systems require extremely low latency to ensure real-time data synchronization and decision-making between vehicles. This low latency reduces the risk of traffic accidents and improves traffic efficiency.
- **Throughput:** Due to the need to process large volumes of vehicle data and sensor information, intelligent transportation systems also have relatively high throughput requirements. The system needs to handle and transmit large amounts of data to ensure smooth and efficient traffic flow.

Xu et al. [1] proposed a new protocol for coordinating autonomous maneuvers through inter-vehicle communication to improve traffic efficiency and reduce conflict risks. They designed a detailed protocol that allows autonomous vehicles to share and negotiate trajectories upon detecting potential conflicts, ensuring all vehicles can adjust their trajectories in a coordinated manner. The protocol comprises three main stages: detection, negotiation, and execution. In the detection stage, each vehicle monitors the surrounding environment via sensors and broadcasts its planned trajectory, enabling other vehicles to anticipate its future movements. During the negotiation stage, if potential conflicts are detected, vehicles renegotiate and propose new trajectory requests, which need to be evaluated and accepted by other vehicles. Finally, in the execution stage, all involved vehicles adjust and execute the new trajectories based on the negotiation results to ensure driving safety and enhance traffic efficiency. The entire process considers the real-time dynamic traffic environment, ensuring efficient and safe vehicle coordination. In simulation tests, their method reduced cumulative time loss by 50% when handling vehicle merging scenarios compared to traditional right-of-way rules, demonstrating the protocol's effectiveness and practicality in real-world applications.

Tseng et al. [2] developed an advanced consensus-based approach in their study to detect and handle faulty information in vehicle platooning. This method includes two stages: fault vehicle detection and controller gain adjustment. First, faulty positional information is identified by analyzing the transmitted data from vehicles. Second, the controller gain is adjusted to mitigate the impact of this faulty information and restore normal platoon behavior. Additionally, they developed a dual-stage mechanism to address faulty speed information from the lead vehicle. Initially, positional information faults are handled through fault detection and controller gain adjustment. In the second stage, faulty speed information from the lead vehicle is addressed by adjusting the speed reference, thereby ensuring the stability and accuracy of the entire platoon. The method also relaxes the assumptions about inter-vehicle communication through an improved CUBA method. Experimental results show that this approach achieves minimal steady-state error and rapid stabilization time in faulty scenarios, outperforming existing technologies.

5.2.2 *Smart Agriculture*

In smart agriculture, wireless consensus has important applications, and these technologies can effectively enhance the reliability and security of agricultural IoT systems. Wireless consensus algorithms ensure that agricultural sensor networks can stably reach consensus in the event of partial node failures or unreliable communications. This is crucial for real-time monitoring of environmental data such as soil moisture and climate conditions, ensuring data consistency and accuracy, and thereby supporting precision agriculture operations like intelligent irrigation and fertilization. Wireless blockchain protocols, through decentralized and distributed ledger technology, provide an efficient and transparent agricultural data management method. Blockchain can record all data collected from sensors and automatically execute agricultural operations through smart contracts. This not only improves data transparency and immutability but also simplifies the management of agricultural supply chains, ensuring that data from the field to the table is authentic and reliable at every step.

I. Network Conditions

- **Common Devices:** In smart agriculture, common devices include environmental sensors (such as temperature, humidity, and light sensors), automated irrigation systems, drones, agricultural robots, and field monitoring equipment. These devices are used to monitor and regulate the crop growth environment and optimize agricultural production processes.
- **Wireless Communication Protocols:** Common wireless communication protocols include LoRa, Zigbee, Wi-Fi, LTE, and 5G. These protocols feature low power consumption and long-distance transmission, suitable for the communication needs of large agricultural fields.
- **Network Architecture:** Smart agriculture typically employs a multi-hop network architecture with a static topology, supported by base stations (such as LTE and 5G) or without base stations (such as LoRa). This network architecture can cover extensive agricultural areas, ensuring stable and efficient communication between devices.

II. Consensus Requirements

- **Latency:** Smart agriculture has moderate latency requirements, needing to ensure timely transmission and processing of environmental data to make real-

time agricultural management decisions. Excessive latency may impact the efficiency and accuracy of agricultural operations.
- **Throughput:** Smart agriculture has moderate throughput requirements, needing to process large amounts of environmental monitoring data and device status information to ensure efficient and precise agricultural operations. The system must have sufficient throughput to handle peak communication demands.

Srbinovska et al. [3] developed a wireless sensor network-based precision agriculture monitoring system for vegetable greenhouses, aimed at reducing management costs and optimizing performance through scientific cultivation. The system monitors key environmental parameters such as temperature, humidity, and light intensity, allowing farmers to adjust crop production strategies in real time. The research team established the sensor network by analyzing and selecting appropriate sensors and wireless technologies. Additionally, they explored different wireless network topologies and communication strategies to further enhance energy efficiency. In their study, wireless consensus algorithms were identified as having significant potential for cross-regional data sharing and verification within the system.

Ojha et al. [4] studied the applications of wireless sensor networks (WSNs) in precision agriculture, irrigation management, and crop health monitoring, with a particular focus on the practical deployment of these technologies globally, especially in India. They conducted a comprehensive analysis of the potential uses of WSNs, including soil moisture monitoring, crop disease prediction, and environmental condition monitoring, and explored the advantages and limitations of different network architectures, sensor types, and communication technologies. In the context of agricultural WSNs, wireless consensus algorithms play a crucial role. For instance, in widespread farmland, multiple sensor nodes need to collaborate to decide whether to activate irrigation systems or adjust fertilization plans. Algorithms like RAFT or Paxos can ensure correct decisions are made even if some nodes become disconnected due to failures or data delays. Additionally, the practical Byzantine fault-tolerant (PBFT) algorithm enhances network stability and robustness in the face of security threats by tolerating a certain proportion of malicious nodes.

5.2.3 Industrial Control Systems

In industrial control systems (ICS), the application prospects of wireless consensus are very promising. Fault-tolerant and Byzantine fault-tolerant wireless consensus ensure that the Industrial Internet of Things (IIoT), such as sensors, controllers, and actuators, can maintain consistent decision-making in the face of communication failures, equipment malfunctions, or malicious attacks. This guarantees the continuity and efficiency of the production process, preventing issues caused by erroneous data. Wireless blockchain technology is widely applied in ICS, particularly in the secure transmission and storage of data. The decentralized and tamper-proof characteristics of blockchain technology ensure the security and trustworthiness of data exchange between industrial devices and between devices and management platforms. For instance, production data and equipment maintenance records can be recorded on the chain in real time, preventing data tampering and providing transparent and reliable industrial information management.

I. Network Conditions

- **Common Devices:** Key devices in industrial control systems include industrial robots, sensors, controllers, and programmable logic controllers (PLCs). Industrial robots execute manufacturing tasks, sensors monitor the environment, and controllers and PLCs command and control the production process.
- **Wireless Communication Protocols:** Common wireless communication protocols include Industrial Ethernet, Wi-Fi, Zigbee, and 5G. Industrial Ethernet and Wi-Fi provide high bandwidth and low latency communication, suitable for real-time control and data transmission.
- **Network Architecture:** Industrial control systems typically use a hybrid network architecture combining single-hop and multi-hop, with a static topology and no base stations. This network architecture provides stable communication, ensuring continuous operation and efficient management of the production line.

II. Consensus Requirements

- **Latency:** Industrial control systems have very high latency requirements, needing extremely low latency to ensure real-time control and efficient operation of the production line. Any delay could cause production line stoppages or decrease product quality.
- **Throughput:** Industrial control systems have moderate throughput requirements, needing to handle sensor data and control instructions to ensure stable and efficient production lines. The system must process large amounts of data to guarantee the smooth operation of the production process.

Kruger et al. [5] studied the application of IPv6 in low-power personal area networks (6LoWPAN), the IPv6 routing protocol (RPL), and the constrained application protocol (CoAP) in industrial wireless sensor networks (IWSN) to validate the feasibility of these technologies in realizing the Internet of Things (IoT). They constructed an experimental environment using the TinyOS-based BLIP 2.0 6LoWPAN stack and TelosB sensor nodes, simulating different network topologies (linear, star, and mesh), and thoroughly tested the effects of communication delay, changes in node distance, transmission power adjustments, and multi-hop communication. Through these experiments, they analyzed the impact of gateway delay and packet size on network performance and provided deployment guidelines for IP-based wireless sensor networks in industrial environments. In such a network environment, wireless consensus algorithms like RAFT are used to ensure data consistency and reliability, particularly in maintaining operational atomicity and system state consistency in the event of node failures or network partitions. These algorithms ensure that critical operational commands are correctly replicated and executed across all key nodes, thus maintaining normal system operation and the accuracy of sensor data in unstable systems.

Aponte-Luis et al. [6] designed a wireless sensor network system based on the IEEE 802.15.4 standard, aimed at achieving remote monitoring and control of industrial parameters through tree-based routing nodes and real-time monitoring applications. The system achieves low-power, high-reliability data transmission through efficient wireless communication and an integrated information management system while incorporating energy management and data encryption technologies to ensure the security and robustness of the system. In the type of systems described in this chapter, wireless consensus algorithms play a crucial role, particularly in ensuring data consistency and system fault tolerance. For example, the RAFT algorithm uses an election mechanism to determine a leader node, optimizing data collection and distribution, thereby ensuring efficient system operation even in the event of partial node failures. On the other hand, the PBFT algorithm enhances the system's resilience to network attacks by performing complex verification processes among nodes, thereby strengthening data authenticity and overall system security.

5.2.4 UAV Networks

In unmanned aerial vehicles (UAV) networks, wireless consensus technology aims to ensure that the network can reach a consistent decision in a wireless communication environment, even in the presence of node failures or malicious nodes. Byzantine fault-tolerant wireless consensus algorithms further enhance fault tolerance, providing higher security for applications of UAVs in security-sensitive missions, such as post-disaster search and rescue and military operations. Wireless blockchain protocols, through distributed ledger technology, offer UAV networks a transparent and tamper-proof data recording mechanism, ensuring the integrity and traceability of mission data. These protocols not only achieve decentralized task management but also effectively prevent data tampering and forgery, making them suitable for fields requiring high data reliability, such as logistics and environmental monitoring.

I. Network Conditions

- **Common Devices:** Key devices in UAV communication include unmanned aerial vehicles (UAVs), ground control stations, and flight sensors. UAVs are used for executing missions and data collection, ground control stations for commanding and controlling the UAVs, and flight sensors for environmental monitoring and data gathering.
- **Wireless Communication Protocols:** Common wireless communication protocols include Wi-Fi, LTE, 5G, and dedicated UAV communication protocols such as Automatic Dependent Surveillance-Broadcast (ADS-B). These protocols feature long-distance transmission and low latency, suitable for communication between UAVs and between UAVs and ground stations.

- **Network Architecture:** UAV communication typically employs a multi-hop network architecture with a dynamic topology, supported by both base stations (such as LTE and 5G) and non-base station networks (such as dedicated UAV communication). This flexible network architecture can adapt to the mobility and mission requirements of UAVs, ensuring continuous and stable communication.

II. Consensus Requirements

- **Latency:** UAV communication requires low latency to ensure real-time coordination and control of UAV formations. High latency could result in mission failure or loss of control over the UAVs.
- **Throughput:** UAV communication requires moderate throughput to handle sensor data from UAVs and control commands, ensuring the smooth execution of missions. The system must be able to process large amounts of data to support real-time decision-making and operations.

Wang et al. [7] proposed blockchain-based frameworks for UAV-assisted disaster rescue information sharing in two separate studies. In the first study, they developed a solution called RescueChain, which employs a lightweight blockchain framework to ensure data sharing and track entities exhibiting abnormal behavior. They also designed a reputation-based consensus protocol to adapt to environments with weak network connections. Furthermore, they utilized ground vehicles as mobile fog nodes, employing an off-chain mechanism based on vehicular fog computing to offload heavy data processing tasks from UAVs and designed a dual-layer reinforcement learning incentive algorithm to promote resource sharing. In the second study, Wang et al. [8] utilized vehicular fog computing (VFC) and mobile edge computing (MEC) technologies, with ground vehicles assisting UAVs in data processing. They proposed a dual-layer resource optimization strategy based on Q-learning and used smart contracts to optimize resource allocation and task offloading. Wireless consensus algorithms such as PBFT, Paxos, and RAFT are crucial in ensuring data consistency and system reliability in disaster environments, especially where networks may be unstable or frequently changing. These algorithms support the construction and maintenance of emergency communication networks in disaster areas, ensuring the timeliness of rescue operations and the accuracy of data.

Kong et al. [9] developed a lightweight asynchronous provable Byzantine fault-tolerant consensus mechanism (LAP-BFT) for unmanned aerial vehicle (UAV) networks. Recognizing the limited resources of UAV networks, they proposed a method to reduce communication overhead by partitioning local trusted state transaction sets and conducting reliable broadcast-controlled transmission (DRBC). They introduced vector commitments to achieve multi-valued Byzantine agreement (PMVBA) for identity and data. This approach not only alleviates the bandwidth and computational burden on individual nodes but also avoids the high energy consumption caused by threshold key operations. Validation through the QualNet simulation platform demonstrated that this method outperforms existing practical asynchronous consensus algorithms in terms of throughput, consensus delay, and energy consumption rate.

5.2.5 Vehicle-to-Everything (V2X)

In vehicle-to-everything (V2X) communication systems, the introduction of wireless consensus and blockchain technology can significantly enhance the system's robustness, security, and data credibility. Fault-tolerant and Byzantine fault-tolerant consensus ensure that vehicles, infrastructure, and other devices in the V2X network can maintain consistent decision-making in the face of communication failures, equipment malfunctions, or malicious attacks. This guarantees the continuity and efficiency of the transportation system, preventing traffic issues caused by erroneous information. Wireless blockchain technology is widely applied in V2X systems, particularly in the secure transmission and storage of data. The decentralized and tamper-proof characteristics of blockchain technology ensure the security and trustworthiness of data exchange between vehicles and between vehicles and infrastructure. For instance, vehicle driving data and traffic accident records can be recorded on the chain in real time, preventing data tampering and providing transparent and reliable traffic information management.

I. Network Conditions

- **Common Devices:** Key devices in V2X include onboard units (OBUs), roadside units (RSUs), and traffic signal controllers. OBUs facilitate communication between vehicles and between vehicles and infrastructure, RSUs manage and relay data,, and traffic signal controllers manage traffic signals.
- **Wireless Communication Protocols:** Common wireless communication protocols include cellular vehicle-to-everything (C-V2X), dedicated short-range communications), and 5G. These protocols feature low latency and high reliability, making them suitable for communication between vehicles and between vehicles and infrastructure.
- **Network Architecture:** V2X systems typically employ a multi-hop network architecture with a dynamic topology, supported by base stations (such as C-V2X). This flexible network architecture can adapt to vehicle mobility and communication needs, ensuring timely data transmission.

II. Consensus Requirements

- **Latency:** V2X requires extremely low latency to ensure real-time communication and safe driving between vehicles. High latency could lead to traffic accidents or disrupt traffic flow.
- **Throughput:** V2X demands high throughput to handle large volumes of vehicle and sensor data, ensuring smooth and efficient traffic. The system must be able to process large amounts of data during peak times to support traffic management and decision-making.

Li et al. [10] studied the coordination issues in connected vehicles under communication and sensor failures by developing a distributed fault-tolerant consensus algorithm to address these problems. This algorithm allows vehicles to achieve coordination without knowledge of the network topology and control systems, and it has low overhead. The proposed ABC algorithm utilizes one-hop communication between vehicles and achieves high fault and message loss tolerance through multiple rounds of state exchanges. Evaluation results show that the algorithm has a low fallback probability under extreme conditions and can quickly reach consensus despite a high message loss rate, effectively improving the overall efficiency and reliability of the system.

Feng et al. [11] proposed a wireless distributed consensus framework for vehicle-to-vehicle communication networks in their study on autonomous vehicles. They designed a three-stage consensus mechanism based on practical Byzantine fault tolerance (PBFT) and introduced a veto collection phase and a gossip phase to address the stringent and complex requirements of vehicle operations. They also proposed a planning tree synthesis to reach a consensus on a series of decisions based on the decision preferences of network members. Simulations demonstrated that this consensus mechanism could achieve and propagate consensus in the network under poor wireless communication conditions, even with faulty vehicles present. Compared to traditional centralized decision-making methods, this scheme significantly enhances the security and efficiency of the system in handling unstable wireless communication environments.

5.2.6 Supply Chain Management

In supply chain management, fault-tolerant wireless consensus, Byzantine fault-tolerant wireless consensus, and wireless blockchain protocols each have their unique applications. Fault-tolerant wireless consensus ensures consistency among nodes within the supply chain, maintaining system stability and continuity, which is particularly important in distributed inventory management, logistics tracking, and real-time inventory monitoring. Byzantine fault-tolerant wireless consensus further enhances system robustness, allowing the system to reach consensus even in the presence of malicious nodes or tampered information, which is crucial for data integrity and security, especially in fraud prevention and ensuring transaction transparency in areas such as cross-border trade and sensitive goods transportation. Wireless blockchain protocols store and manage information from various stages of the supply chain in a distributed manner, ensuring data transparency and traceability, improving supply chain visibility and efficiency, and reducing operational risks. For example, in the food supply chain, blockchain technology can be used to track the origin, processing, and transportation of products, ensuring food safety.

I. Network Conditions

- **Common Devices:** Key devices in supply chain management include Radio Frequency Identification (RFID) readers, sensors, mobile terminals, and servers. RFID readers are used for tracking items, sensors for monitoring environmental conditions, mobile terminals for data collection and processing, and servers for data storage and management.

- **Wireless Communication Protocols:** Common wireless communication protocols include Wi-Fi, LoRa, LTE, and 5G. These protocols feature long-distance transmission and low power consumption, suitable for the various communication needs in supply chain management.
- **Network Architecture:** Supply chain management systems typically use a multi-hop network architecture with a dynamic topology and base stations. This network architecture can cover extensive areas, ensuring stable and efficient communication between different stages of the supply chain.

II. Consensus Requirements

- **Latency:** Supply chain management has moderate latency requirements, needing to ensure that data from various stages of the supply chain is updated and verified in a timely manner. Excessive latency could lead to information lag, affecting the efficiency and transparency of the supply chain.
- **Throughput:** Supply chain management has high throughput requirements, needing to handle large volumes of tracking data to ensure transparency and efficient operation of the supply chain. The system must be capable of processing large amounts of data to support real-time tracking and management.

Awan et al. [12] proposed a future smart agriculture and food supply chain model combining the Internet of Things (IoT) and blockchain technology. They designed an architecture comprising a physical data layer, a logical data layer, and a web interface layer. IoT nodes monitor the farm environment and crop growth, and the data is transmitted to the database via wireless routers and gateways. To enhance energy efficiency, they developed a new clustering protocol based on the LEACH protocol. This protocol improves energy efficiency and network stability by staging data transmission among IoT nodes. Additionally, they integrated blockchain technology with IoT, using smart contracts for data exchange within the system to ensure the traceability and safety of food in the supply chain. Simulation results showed that the new protocol improved network stability by 23%, reduced energy consumption by 68%, and extended network life by 112% compared to the LEACH protocol.

Jangirala et al. [13] proposed LBRAPS, an efficient and lightweight blockchain-based RFID authentication protocol designed for supply chains in a 5G mobile edge computing environment. This protocol ensures efficiency and security through the use of bitwise XOR, one-way encryption hash, and bitwise rotation operations. The protocol consists of two phases: initialization and authentication key agreement, effectively defending against various attacks. The researchers used the AVISPA tool for simulation verification, demonstrating that LBRAPS has superior security compared to existing protocols. Experimental results showed that LBRAPS achieved a better balance between security and functionality, as well as communication and computation costs.

5.2.7 *Smart Healthcare*

In smart healthcare systems, through fault-tolerant and Byzantine fault-tolerant consensus, various devices in the medical Internet of Things (such as monitors, diagnostic equipment, and wearable devices) can maintain consistent decision-making in the face of communication failures, equipment malfunctions, or malicious attacks. This ensures the continuity and accuracy of medical services, preventing erroneous data from affecting patient treatment. Wireless blockchain technology is widely applied in smart healthcare, particularly in the secure transmission and storage of data. The decentralized and tamper-proof characteristics of blockchain technology ensure the security and trustworthiness of data exchange between medical devices and between devices and management platforms. For instance, patient medical records and treatment records can be recorded on the chain in real time, preventing data tampering and providing transparent and reliable medical information management.

I. Network Conditions

- **Common Devices:** Key devices in medical health records include medical sensors, mobile terminals, servers, and electronic health record (EHR) systems. Medical sensors monitor patients' physiological data, mobile terminals are used for data collection and transmission, servers store and process medical data, and EHR systems manage and access patient records.
- **Wireless Communication Protocols:** Common wireless communication protocols include Wi-Fi, Bluetooth, LTE, and 5G. These protocols feature high bandwidth and high reliability, suitable for the transmission and management of medical data.
- **Network Architecture:** Medical health record systems typically use a single-hop network architecture with a static topology and base stations. This network architecture provides stable communication, ensuring the secure and efficient management of medical data.

II. Consensus Requirements

- **Latency:** Medical health records require low latency to ensure the quick recording and access of medical data, supporting real-time monitoring and emergency treatment. High latency could affect patient treatment outcomes and the quality of healthcare services.
- **Throughput:** Medical health records require moderate throughput to handle large volumes of patient data, ensuring data integrity and security. The system must be capable of processing substantial amounts of data to support the real-time updating and management of medical records.

Jeong et al. [14] researched an intelligent medical monitoring system that leverages blockchain technology combined with the Internet of Things (IoT) to enhance the reliability and security of personal medical information management. They measured individuals' electrocardiogram (ECG) information through multiple sensors and used blockchain technology for data storage and real-time monitoring. In their study, they designed and implemented a monitoring system that collects biosignals via sensor units, processes these signals through control units, and provides real-time information to users, caregivers, and medical experts via smartphones. Experimental results showed that the system achieved a 19.2% improvement in storage efficiency and classification accuracy of up to 97.2%.

Pham et al. [15] proposed a blockchain smart contract-based telemedicine system aimed at protecting the personal and device-generated information of patients. They used the Ethereum protocol to create a telemedicine system that includes hospitals, doctors, and patients. Sensors measure patients' health status, and the data is automatically written to the blockchain. To efficiently store medical device information, they proposed a data processing mechanism to filter sensor data, determining whether to write the data to the blockchain, thus reducing blockchain size and transaction costs. When abnormal data is detected, the system immediately writes it to the blockchain and triggers an emergency mechanism to contact doctors and hospitals. The system's effectiveness was validated in the Ethereum test environment and implemented in a small-scale experimental setup. Comparisons showed that using the blockchain system provided superior data privacy, availability, immutability, and transparency compared to traditional systems.

5.3 Summary

This chapter mainly summarizes the application scenarios from Chaps. 2, 3, and 4. No single consensus algorithm can be suitable for all application scenarios. Compared to wired networks, the most significant characteristic of wireless networks is their greater diversity, primarily reflected in communication devices, communication protocols, and network architectures. Therefore, this diversity is particularly prominent in wireless networks compared to wired networks. We introduce the requirements of three types of wireless consensus protocols for application scenarios, focusing mainly on communication and computation needs. Then, we present the seven most important application scenarios for wireless consensus. For each application scenario, we describe the basic network conditions and the consensus requirements. The basic network conditions include three aspects: communication devices, communication protocols, and network architectures. The consensus requirements mainly include two key metrics: latency and throughput.

References

1. Xu, W., Willecke, A., Wegner, M., Wolf, L., & Kapitza, R. (2019). Autonomous maneuver coordination via vehicular communication. In *2019 49th Annual IEEE/IFIP International Conference on Dependable Systems and Networks Workshops (DSN-W)* (pp. 70–77). IEEE.

2. Tseng, T.-Y., Huang, D.-J., Lin, J.-Y., Chang, P.-J., Lin, C.-W., & Liu, C. (2023). Consensus-based fault-tolerant platooning for connected and autonomous vehicles. In *2023 IEEE Intelligent Vehicles Symposium (IV)* (pp. 1–8). IEEE.
3. Srbinovska, M., Gavrovski, C., Dimcev, V., Krkoleva, A., & Borozan, V. (2015. Environmental parameters monitoring in precision agriculture using wireless sensor networks. *Journal of Cleaner Production, 88*, 297–307.
4. Ojha, T., Misra, S., & Raghuwanshi, N. S. (2015). Wireless sensor networks for agriculture: The state-of-the-art in practice and future challenges. *Computers and Electronics in Agriculture, 118*, 66–84.
5. Kruger, C. P., & Hancke, G. P. (2014). Implementing the internet of things vision in industrial wireless sensor networks. In *2014 12th IEEE International Conference on Industrial Informatics (INDIN)* (pp. 627–632). IEEE.
6. Aponte-Luis, J., Gómez-Galán, J. A., Gómez-Bravo, F., Sánchez-Raya, M., Alcina-Espigado, J., & Teixido-Rovira, P. M. (2018). An efficient wireless sensor network for industrial monitoring and control. *Sensors, 18*(1), 182.
7. Wang, Y., Su, Z., Xu, Q., Li, R., & Luan, T. H. (2021). Lifesaving with rescuechain: Energy-efficient and partition-tolerant blockchain based secure information sharing for uav-aided disaster rescue. In *IEEE INFOCOM 2021-IEEE Conference on Computer Communications* (pp. 1–10). IEEE.
8. Wang, Y., Su, Z., Xu, Q., Li, R., Luan, T. H., & Wang, P. (2023). A secure and intelligent data sharing scheme for uav-assisted disaster rescue. *IEEE/ACM Transactions on Networking, 31*(6), 2422–2438.
9. Kong, L., Chen, B., & Hu, F. (2022). Lap-bft: Lightweight asynchronous provable byzantine fault-tolerant consensus mechanism for uav network trusted systems. In *Asia-Pacific Web (APWeb) and Web-Age Information Management (WAIM) Joint International Conference on Web and Big Data* (pp. 232–246). Springer.
10. Li, T., Tseng, L., Higuchi, T., Ucar, S., & Altintas, O. (2021). Poster: Fault-tolerant consensus for connected vehicles: A case study. In *2021 IEEE Vehicular Networking Conference (VNC)* (pp. 133–134). IEEE.
11. Feng, C., Xu, Z., Zhu, X., Klaine, P. V., & Zhang, L. (2023). Wireless distributed consensus in vehicle to vehicle networks for autonomous driving. *IEEE Transactions on Vehicular Technology, 72*(6), 8061–8073.
12. Awan, S., Ahmed, S., Ullah, F., Nawaz, A., Khan, A., Uddin, M. I., Alharbi, A., Alosaimi, W., & Alyami, H. (2021). Iot with blockchain: A futuristic approach in agriculture and food supply chain. *Wireless Communications and Mobile Computing, 2021*, 1–14.
13. Jangirala, S., Das, A. K., & Vasilakos, A. V. (2019). Designing secure lightweight blockchain-enabled rfid-based authentication protocol for supply chains in 5g mobile edge computing environment. *IEEE Transactions on Industrial Informatics, 16*(11), 7081–7093.
14. Jeong, S., Shen, J.-H., & Ahn, B. (2021). A study on smart healthcare monitoring using iot based on blockchain. *Wireless Communications and Mobile Computing, 2021*, 1–9.
15. Pham, H. L., Tran, T. H., & Nakashima, Y. (2018). A secure remote healthcare system for hospital using blockchain smart contract. In *2018 IEEE Globecom Workshops (GC Wkshps)* (pp. 1–6). IEEE.

SPRINGER NATURE

GPSR Compliance

The European Union's (EU) General Product Safety Regulation (GPSR) is a set of rules that requires consumer products to be safe and our obligations to ensure this.

If you have any concerns about our products, you can contact us on ProductSafety@springernature.com

In case Publisher is established outside the EU, the EU authorized representative is:

Springer Nature Customer Service Center GmbH
Europaplatz 3
69115 Heidelberg, Germany

The manufacturer's authorised representative in the EU is Springer Nature Customer Service Centre GmbH, Europaplatz 3, 69115 Heidelberg, Germany. If you have any concerns regarding our products, please contact ProductSafety@springernature.com

Printed and bound by CPI Group (UK) Ltd, Croydon, CR0 4YY
01/10/2026
02225960-0003